VOLCANOES

VOLCANOES

WHAT EVERYONE NEEDS TO KNOW®

TAMIE J. JOVANELLY

OXFORD
UNIVERSITY PRESS

OXFORD
UNIVERSITY PRESS

Oxford University Press is a department of the University of Oxford. It furthers the University's objective of excellence in research, scholarship, and education by publishing worldwide. Oxford is a registered trade mark of Oxford University Press in the UK and certain other countries.

Published in the United States of America by Oxford University Press
198 Madison Avenue, New York, NY 10016, United States of America.

CIP data is on file at the Library of Congress

ISBN 978–0–19–781747–6 (pbk.)
ISBN 978–0–19–781746–9 (hbk.)

DOI: 10.1093/wentk/9780197817469.001.0001

Paperback printed by Sheridan Books, Inc., United States of America
Hardback printed by Lightning Source, Inc., United States of America

The manufacturer's authorized representative in the EU for product safety is Oxford University Press España S.A. of Parque Empresarial San Fernando de Henares, Avenida de Castilla, 2 – 28830 Madrid (www.oup.es/en or product.safety@oup.com). OUP España S.A. also acts as importer into Spain of products made by the manufacturer.

FSC www.fsc.org
MIX
Paper | Supporting responsible forestry
FSC® C008955

If you have ever had the chance to meet my husband, you would agree that I've won the lottery. The unimaginable things that I have seen and done in the past two decades result from his unwavering excitement to travel the world and to dream big. This book is dedicated to him, Joe Cook.

CONTENTS

INTERPRETING THE LANDSCAPE: VOLCANO GEOMORPHOLOGY PART I

INTERPRETING THE LANDSCAPE: VOLCANO GEOMORPHOLOGY PART II

EROSION

ENERGY

CLIMATE CHANGE

PREDICTION

SUPERVOLCANO

ANCIENT ERUPTIONS

EFFECTS OF VOLCANIC ACTIVITY ON HUMAN LIFE

PLANETARY GEOLOGY

PREFACE

My first experience with a volcano occurred when I was a Doctoral Fellow at the University of Tokyo Earthquake Research Institute, where I held the position of Visiting Scientist, studying tsunamis. When I had free time on the weekends, I would take solo trips on the bullet train to explore the country. At the first chance, I headed to Mt. Fuji. As any inexperienced hiker would be, I was largely unprepared for the overnight ascent of a 3,776 m volcano. I bought cheap rain gear and men's boots that were too big, and yet I met the climb with enthusiasm for an iconic geologic feature that I knew just enough about to keep me in awe without having any fear of being on her flanks. I did not speak Japanese, but that didn't matter. Hiking this volcano is intended to be a profoundly quiet and an introspective spiritual pilgrimage, one that is meant to be completed only once in a lifetime. The hike started at 8:00 p.m. The evening starting time is planned so that the summit can be reached just as the sun is rising. It was cold and rainy, and I quickly realized that I was going to be miserable for the next 24 hours. I distinctly remember reading the Lonely Planet guidebook on my way to the trailhead, and it said that the hike was "so easy your grandmother could do it." While there were an incredible number of grandmothers at the trailhead that night, the challenging ascent that included bouldering up basalt piles, around massive lava bombs, and through slippery, unconsolidated tephra made me feel like this was a big adventure that would change my life. And it would.

The Mt. Fuji pilgrimage solidified my love for the climb, and I became a volcano chaser. I made it a goal to visit a new volcanic region every year, either through my summer research as a hydrologist, leading international study abroad programs as a geology professor, or through other travel. As my curiosity and understanding of volcanology continued to grow throughout my career in geology, the focus of the ascent shifted from summit-bagging to appreciating the structure, process, and time involved in forming a volcano. The more volcanoes I visited, the more I was able to make interesting comparisons. For example, the shape of Concepción (Nicaragua) to that of Arenal (Costa Rica), the crater of Mount Vesuvius (Italy) against Askja (Iceland), the glacial influence at Mount Rainer (Washington, USA) to Mount Kilimanjaro (Tanzania, Africa), or the active volcanism I witnessed at Stromboli (Italy) to Fagradalsfjall (Iceland).

With that stated, this book is the result of long, slow, quiet, upward hikes on volcanic flanks that provided me with extended periods of time to contemplate volcanic processes as I stepped one foot in front of the other, offering a happy alternative to complaining. And the content of this text was inspired by nearly two decades of teaching about volcanoes, both to inquisitive undergraduate students and to the guests who attend geologic tours with my company Adventure Geology Tours. In fact, much of the subject matter contained in this book originated as essay questions found on my exams, or should I say "grueling" exams, to quote my students.

The readers of *Volcanoes: What Everyone Needs to Know* will follow a similar journey, approaching the book first as an enthusiastic novice and ending with an in-depth understanding of volcanoes that accumulates with each chapter. Here, I want the readers to know how much FUN I had writing this book (*a literal blast*). I hope that the approachable question-and-answer format makes the material easy to process while sparking continued curiosity that progresses in complexity. Within the book, I present some necessary introductory geology concepts used to interpret volcanic landscapes (Chapters 1 through 4) before delving into specific topics such as volcanoes as natural hazards (Chapter 5), how volcanic landscapes erode (Chapter 6), how geothermal energy is harnessed (Chapter 7), the impact of volcanic eruptions on climate (Chapter 8), and concepts surrounding volcanic prediction and prevention (Chapter 9). The remaining chapters highlight the most significant volcanic eruptions on the planet (Chapter 10) and their overall effects on human life and civilization (Chapters 11 and 12) before concluding with compelling ideas regarding planetary volcanism, such as the potential for volcanic activity on other planets and moons in our solar system (Chapter 13). Also included within the book are more than 60 color diagrams and illustrations to clarify complex concepts, GPS coordinates (latitude/longitude) for easy reference to volcanoes on virtual platforms, color photographs of common volcanic minerals and rocks useful for identification, a glossary of geological terms related to volcanism, and an appendix featuring 100 of the world's most active volcanoes. Although the book is intended to be read from cover to cover, a traveler or a student may choose to use the index to search for specific events, processes, or volcanoes of interest. Likewise, the photographs of common volcanic minerals and rocks found in Chapter 2 can be used as a reference to identify samples.

ABOUT THE AUTHOR

Volcanoes: What Everyone Needs To Know® is Dr. Tamie J. Jovanelly's second book. This book follows her best-selling monograph published by Wiley in 2020 for the American Geophysical Union titled *Iceland: Tectonics, Volcanics, and Glacial Features*. To find out more information about the author, visit www.volcanoeswhateveryoneneedstoknow.com, or her tour company webpage, www.adventuregeologytours.com.

ACKNOWLEDGMENTS

The only other person I know who loves volcanoes as much as I do is Thomas Griffith. I thank him for his equal enthusiasm, which led us down many geological rabbit holes (*also known as lava tubes*) during the writing of this book, striking us both with awe and inspiration about how fascinating this subject is. I think he would agree that if there is any profession better than being a geologist, we haven't found it.

Next, my Icelandic friends and colleagues, Gróa Hreinsdóttir and Harpa Sól Gudmundsdóttir, have helped make the volcanic country I love the most feel like my home away from home. I will be forever grateful to them for allowing me to become fully immersed in Icelandic culture through their family gatherings, home-cooked meals, and choir music.

I would also like to recognize the Association for Women Geoscientists (AWG). This international organization, devoted to enhancing the quality and level of participation of women in geosciences, has been a pillar of my academic and professional career. Without a doubt, the innumerable mentors I have gathered over time within the AWG network have helped to direct, support, and craft me into the geoscientist I am today.

BACKGROUND GEOLOGY AND UNDERSTANDING PLATE TECTONICS

1

WHERE DOES LAVA COME FROM, AND WHERE ARE VOLCANOES FOUND ON EARTH?

What are the parts of the Earth?

The Earth, which has a radius of 6,371 km, is a planet that is differentiated or separated into distinct parts, each having unique physical properties relating to changes in temperature, pressure, and depth (Figure 1.1). The invention of the seismograph in 1875 allowed scientists to instrumentally monitor how earthquake-generated seismic waves travel throughout the Earth. With this data, 19th-century seismologists Andrija Mohorovicic and Inge Lehmann theorized with some confidence that the Earth was not simply a solid rock. This was later confirmed as technology improved with time to become a formalized method called seismic tomography. This method uses thousands of earthquakes recorded at seismic stations around the globe to measure the speed of seismic waves traveling through the Earth, allowing geophysicists to make 3-dimensional profiles outlining the structure of the subsurface.

Starting at the surface of the Earth, the crust is the first layer that can be described. The crust, underlying the oceans and continents, is a rigid surface interpreted by its rock type, density, and thickness. The oceanic crust is generalized as basaltic, a kind of extrusive igneous rock with a 3.3 g/cm^3 density.

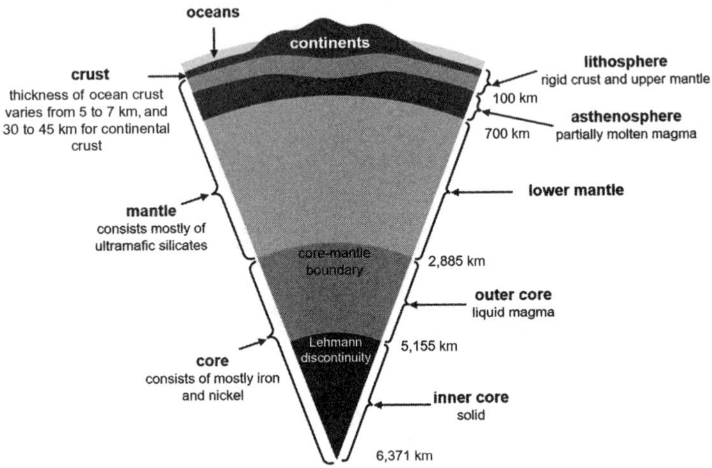

Figure 1.1 Labeled parts of the Earth from surface to inner core.

The oceanic crust has an overall thickness of between 5 and 7 km, which allows it to be pliable at plate boundaries, especially in subduction zones. In comparison, continental crust is characterized as granitic, a type of intrusive igneous rock with a 2.7 g/cm^3 density. The continental crust is considerably thicker than the oceanic crust (35–40 km) due to mountain building and the compounding effects of collisional plate tectonic cycles over the past 4.6 billion years (Figure 1.2).

Combined, the oceanic and continental crusts rest on top of the upper mantle to form the lithosphere. The upper mantle extends from the crust to a depth of 410 km below the surface. Decompression melting occurs within the upper mantle within the solid rock that is held at approximately the same temperature. Still, the pressure is reduced, which encourages the formation of magma (Figure 1.3). Decompression melting creates convection cells, or density-driven systems, that lift magma toward the surface. Thus, within the upper mantle, the solid material that makes up the lithosphere melts into molten magma that can become visible at the surface during volcanic eruptions.

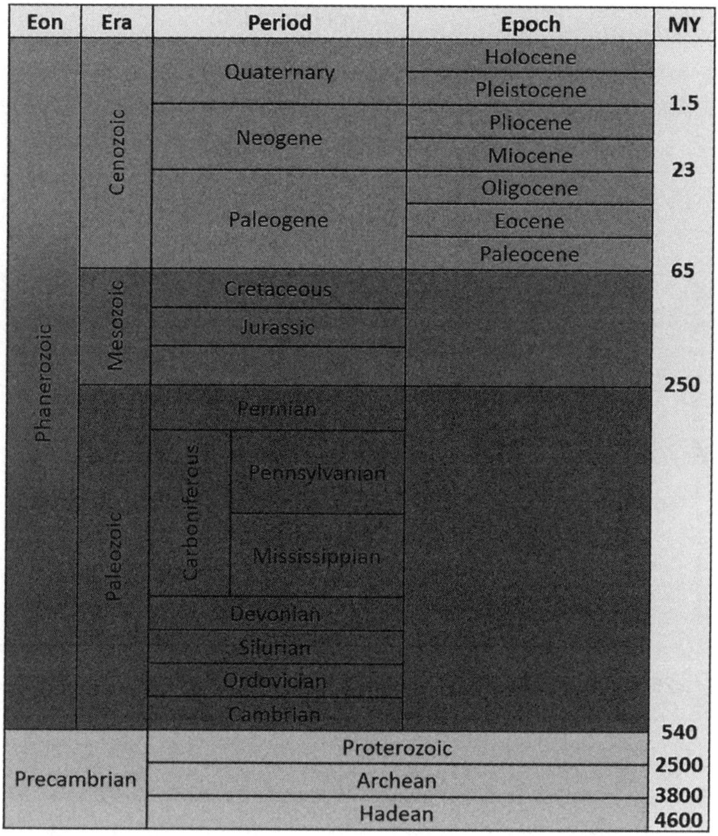

Figure 1.2 The geological time scale with associated ages.

Where the upper mantle ends, the lower mantle begins. This area within the Earth is termed the "asthenosphere." The lithosphere and asthenosphere are separated by the Mohorovičić Discontinuity Line, more commonly referred to as simply the Moho Line. The Moho Line lies about 35 km below the continental crust and about 7 km beneath the oceanic crust. This boundary marks a distinct difference in density between the two systems as described by the rate at which seismic waves can move through each system (i.e., seismic waves move slower

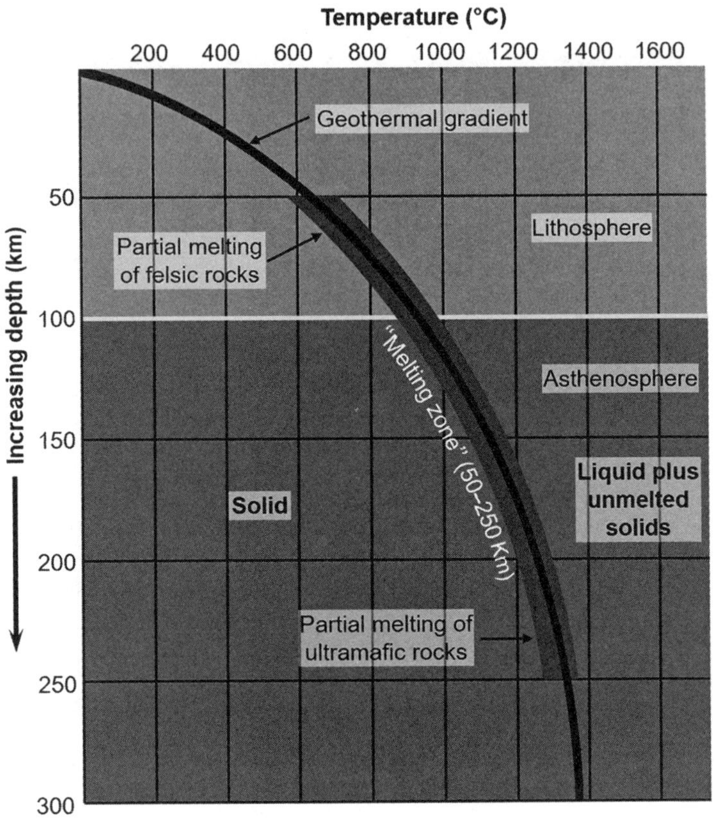

Figure 1.3 Temperature and pressure gradient as compared to lithosphere and asthenosphere.

through the rigid lithosphere and quicken as they meet the less dense asthenosphere). Because the lower mantle is deeper within the Earth, it is under increasingly high temperatures (1,690°C to 2,360°C) and pressure (~1.3 million times atmospheric pressure), creating a molten or plastic magma zone.

The lower mantle ends 2,885 km below the surface to meet the mantle–core boundary. This boundary, referred to as the Lehmann Discontinuity Line, is unique because it is the transitional zone from plastic magma in the lower mantle to fluid magma in the outer core. Composed of mostly iron and nickel, the outer core is in a liquid phase because it is hotter than the mantle. However, just below, the inner core is solid iron and nickel. This results from intense pressure that keeps the inner core from liquifying.

What are the physical properties of the mantle?

In order to understand the physical properties of the mantle, the formation of the Earth must first be understood. Circa 5 BYA, the Earth began to organize itself as a protoplanet as it underwent internal melting to eventually differentiate into distinct layers having specific properties (Figure 1.1). By about 4.5 BYA, the Earth had arranged into a solid outer layer called the crust: the plastic magmatic mantle, and other elements, like nickel and iron, would separate to form the solid inner core. A period of volcanic outgassing occurred from roughly 3.8 to 2.8 BYA, allowing an abundance of water trapped in the mantle to escape to the surface. It is thought that this period of 1 billion years helped to create and stabilize our oxygenated planet. Today, water still exists in the mantle and is returned by subduction processes involved with plate tectonics. Extensive research has revealed that water stored in the mantle is equivalent to 1% of the mass of our oceans, with the majority trapped in the transition zone between the crust and upper mantle (Ohtani, 2020). We see evidence of this water at the surface when volcanoes erupt, releasing water vapor. Geologists refer to this water as "juvenile water," because it is derived from material deep within the Earth and has not previously appeared at the surface. Juvenile water from volcanic eruptions is a new contribution to the Earth's water budget.

The mantle, which represents 84% of the Earth's total volume and 65% of its mass, is the most important layer to generate that surface through tectonic processes or as inclusions in explosive ejectiles (Figure 1.1). The mantle is composed of a group of minerals called silicates. Silicates, or SiO_4^{2-}, are considered to be the building blocks of 95% of the minerals found within the Earth's crust because they can bond to most elements on the periodic table. The most common silicate minerals found within the mantle are olivine, pyroxene, and garnet. More specifically, a variation of olivine called peridotite $(Mg, Fe)_2SiO_4$ is rarely found at the surface. Still, it is a dominant rock type found in the mantle that can be forced to the surface from depth during explosive eruptions. Other elements abundant in the mantle include magnesium oxide and aluminum. The concentration of these additives helps to explain the origin and chemical composition of Mid-Ocean Ridge Basalts (MORB) that line the ocean floor, which are sourced from the upper-mantle region.

The mantle is unique because, unlike the other parts of the Earth, it undergoes multiple viscous transitions. The rigid upper mantle begins at the bottom of the Earth's crust: 7 km for oceanic and 40 km for continental. Rocks that contain high amounts of silica (termed felsic) begin to melt around 50 km below the surface, or 600°C (Figure 1.3). The transition into complete melt occurs around 100 km deep, marking the boundary between the lithosphere and the asthenosphere, known as the Moho Line. Rocks that contain less silica (termed mafic) remain crystalline longer, as they have higher melting temperatures noting that the geothermal gradient increases with depth. For this reason, magma flow does not occur throughout the entire mantle. Instead, most of the mantle is solid rock undergoing solid-state convection, yet it explains the full spectrum of magma generation in the subsurface.

Why doesn't the crust (lithosphere) sink into the mantle (asthenosphere)?

The lithosphere contains rigid oceanic and continental crust and upper mantle, while the asthenosphere contains the lower mantle. Although rigid, the lithosphere does not sink into the asthenosphere for two reasons. First, the lithosphere is less dense than the asthenosphere based on the mathematical equation density = mass/volume, where in Figure 1.1 the differences in volume are shown. That is, the asthenosphere has three times greater mass and volume, which supports the lithospheric layers. Next, the consistency of the lower mantle becomes plastic in texture because, with depth, temperature increases at a rate of 30°C per km (away from plate boundaries). Within the lower mantle, convection currents organize to move hot rising material toward the lithosphere boundary as the colder material sinks. The uplifting circulatory motion lends buoyancy, keeping the lithosphere afloat at the surface.

How is a volcano defined?

A volcano can be formed where gaseous molten magma stored in a subsurface magma chamber rises through a central conduit or cracks in the lithosphere (Figure 1.4). While magma

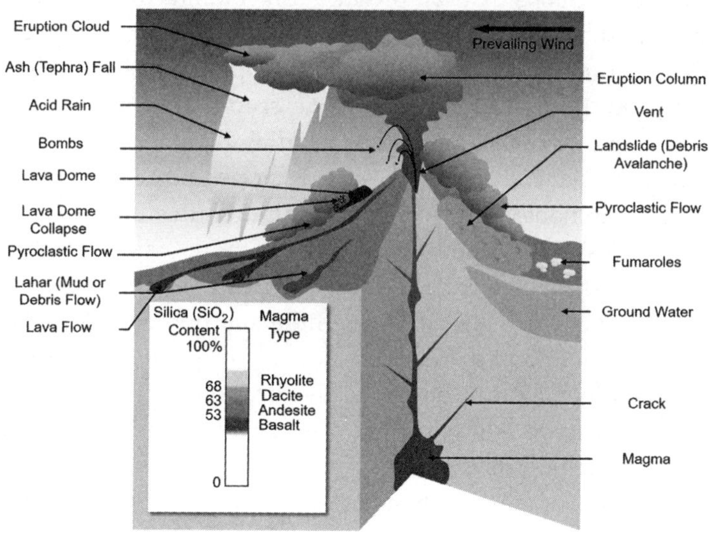

Figure 1.4 Cross section of a volcano with labeled parts. Modified from Stovall et al. (2019).

chambers can be as close as 1 km below the surface, the optimal depth is between 6 and 10 km due to the ability of water vapor to exsolve out of magma and the flexibility of the crust to expand to accommodate chamber growth. When magma solidifies at the Earth's surface, it is called lava.

Volcanoes have a conical shape that can vary in steepness from 5° to 10°, as with a shield volcano, or 30° to 35°, as with a stratovolcano. Volcanoes can be associated with specific landscape features, including vents, craters, lava domes, and fumaroles. Likewise, volcanoes can produce debris products such as ash clouds, lahars, landslides, pyroclastic flows, and tephra. Due to the hot magma stored below the surface, it is not uncommon for nearby hot springs, boiling mud pots, or geysers to form. Additionally, where groundwater is present, it is naturally heated and can be extracted to produce geothermal energy.

The mineralogy of the magma created in the upper mantle dictates a volcano's shape and explosivity. A magma that contains >50% silica is termed felsic and is associated with explosive eruptions that can produce ash columns that can enter into the stratosphere (>10 km above the surface), as seen in the 1991 eruption of Mount Pinatubo, Philippines (939 m; 15.1413, 120.3558). Rocks formed from felsic eruptions are commonly light in color (i.e., pink, white), including rhyolite and pumice. A magma that contains <50% silica is termed mafic and is associated with nonexplosive or gentle eruptions that produce little volcanic ash, like Kilauea, Hawaii (1,031 m; 19.4055, −155.2838). Mafic rocks that are dark in color (i.e., gray to black, green, red) because they are higher in magnesium and iron often form basalt, peridotite, and scoria.

Like other natural disasters, the frequency of a volcano's eruption can be approximated. This is called the repose of a volcano. Geologists can use seismology to understand the size and volume of magma chambers and how quickly the magma will infill the chamber. For some volcanoes, the caldera's size is roughly equal to the size of the magma chamber. A drastic comparison could be the Yellowstone Caldera, Wyoming (2,605 m; 44.4305, −110.7211), roughly 80 km long and 65 km wide.

Before an eruption, moving magma triggers hundreds to thousands of low-energy earthquakes called earthquake swarms. Additionally, as the magma pushes up the vent, the ground surface inflates to accommodate the new buoyant mass. For volcanoes that produce ash, geologists can study the layers deposited around the volcano to understand how many eruptions have happened in the past. This field of science is referred to as tephrachronology.

In areas where underground magma stored in a volcano is not connected to the surface, the magma will undergo cooling and solidification to form coarse-grained plutons. An example of this would be Mount Rushmore National Monument, South Dakota (43.8811, −103.4511), which includes the faces of four presidents carved into the Precambrian pluton called the Harney Peak Granite.

Can a volcano form anywhere?

Volcanoes, which result from magma erupting on the Earth's surface, are formed by two different processes: shifting of tectonic plates or mantle plume (hot spot) upwelling. Plate tectonics is a theory that the Earth's lithosphere is divided into several large, platelike sections that move as distinct landmasses (Figure 1.5). Today, there are seven major plates and eight minor plates. The most enormous plates include the Antarctic, Eurasian, and North American. Although rigid, the lithospheric plates are not stationary. Instead, due to the continual movement of lower mantle convection cells, the plates systematically converge (come together) or diverge (move apart). Cyclically, every 500 MY, the crust that forms continental land masses will collide to create a supercontinent called Pangea.

While volcanoes can form at either convergent or divergent plate boundaries, the most destructive ones are found as plates

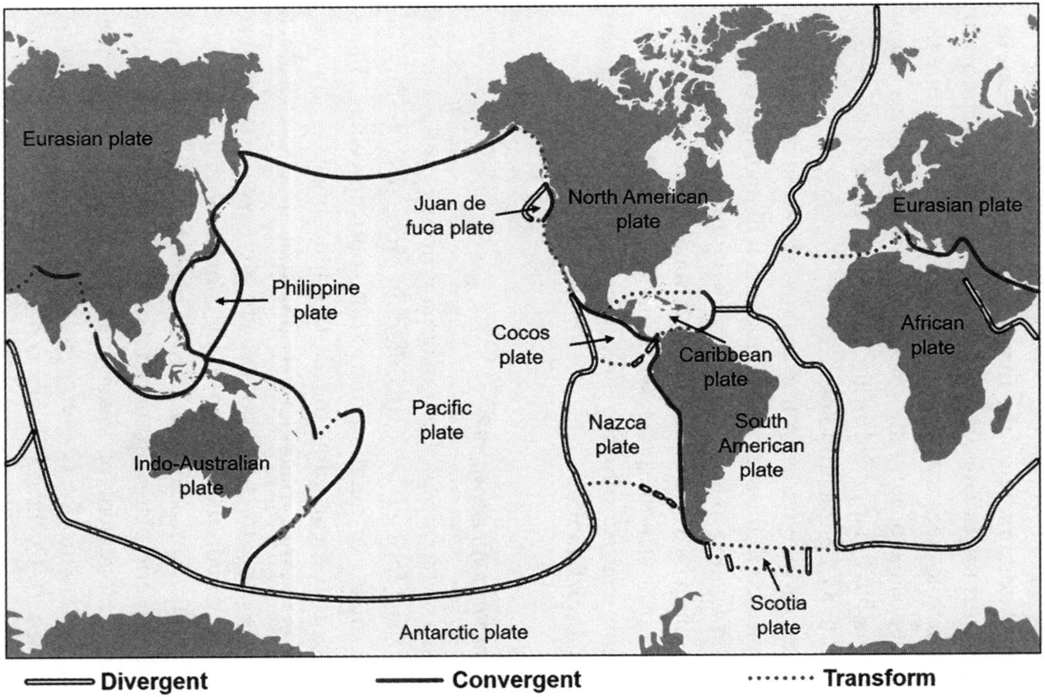

Figure 1.5 Plate tectonic diagram locating global convergent, divergent, and transform boundaries.

are pushed together to form subduction zones. A subduction zone is a collision between two of Earth's tectonic plates, where one plate sinks into the mantle underneath the other plate, forming a trench (Figure 1.6). Convergent plate boundaries forming subduction zones are described by the types of lithospheric crust (oceanic or continental) that are colliding to result in ocean-continent convergence or ocean-ocean convergence. In the first scenario, the oceanic plate made of basalt has a greater density and will subduct underneath the lighter granitic continental plate. Next, suppose the plates are composed of the same oceanic basalt rock type. In that case, the older plate is usually denser and will be the subducting plate. Conversely, when two plates of continental land mass collide, as in the Himalayas, they will rise up as they collide because they are less dense than the underlying lithosphere, creating tall mountains instead of a subduction zone. Without a subduction zone, volcanoes cannot form.

In locations where oceanic-continental crust converges, a continental volcanic arc can form, like the Andean Volcanic

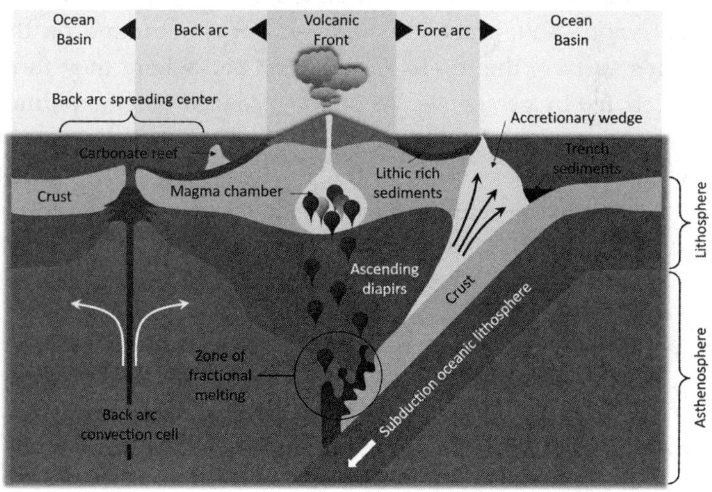

Figure 1.6 Example of subduction zone at ocean–continent convergent plate margin.

Belt along the west coast of South America from Argentina to Peru. Similarly, where oceanic-oceanic crust convergence occurs, a volcanic arc, a chain of volcanic islands that parallels a trench, can form, like the Aleutians Islands (Alaska). In both scenarios, more magma can be produced when ocean water is present at the subduction zone because water lowers the melting temperature of the solid rock being subducted. This process is referred to as flux melting. The abundance of magma at subduction zones creates a specific type of volcano called a stratovolcano. Additionally, water in magma creates a more significant potential for explosive eruptions.

A divergent plate boundary, called the Mid Ocean Ridge, is found at the bottom of oceans worldwide. Here, two plates are pulled apart to create a rift, or conduit, for magma to pile onto the ocean floor that can build a submarine volcano. These volcanoes are usually hidden from view at or above sea level, as they are thousands of meters below the ocean surface. This type of volcanic activity is attributed with making large amounts of new oceanic crust. It accounts for approximately 75% of all effusive eruptions annually. Volcanoes can also be formed by the upwelling of magma within a plate, termed a hot spot. While there are some 50 hot spots found on the Earth's surface, the mechanics of how and where they form are debated by geologists. When the localized magma plumes breach the surface, they create a volcano. If a hot spot occurs in the ocean, a shield volcano will typically result. Realizing that hot spots remain stationary while plates move over the top is essential. This can result in a chain of volcanic islands, like the Galapagos Islands. The oldest islands are located the furthest away from the hot spot. They are no longer volcanically active as they have been transported away from the magma source.

Why are some islands called volcanoes?

By definition, an island is a landmass that is surrounded by water. The planet has innumerable islands, but how they form can be quite different. There are four significant kinds of islands: continental, barrier, coral, and volcanic. Continental islands form from remnant pieces of continental crust that broke off during Pangea cycles, barrier islands form from sediment deposition along a coastline, and coral islands form on shallow ocean floors that allow coral reefs to adhere and grow. Volcanic islands are the only type involving plate tectonics, created from the influence of subduction zones or the presence of a hot spot.

Volcanic island arcs align themselves parallel to subduction zones. There are three prominent volcanic island arcs on the planet that share similar tectonic characteristics: Kuril Islands (Russia; 47.1541, 150.9702), Aleutian Islands (Alaska; 54.8316, −165.8944), and Sunda Arc (Indonesia; 0.9751, 100.9444). Within each of these systems, the angle of the subducting plate is 45°. This, and the consequences of a spherical planet, create a crescent alignment around a radius with a curvature of 20°–22°. As long as active subduction and magma production exist, all islands within the alignment are considered volcanically active. Many times, volcanic arcs border seas, many of which form small ocean basins between island arcs and continental margins.

Volcanic islands can also be created by the upwelling of localized magma under the oceanic crust called a mantle plume or a hot spot. About 50 hot spots have been identified, randomly scattered under the ocean floor and continents. Hot spots vary in height from 500 km to 1,000 km and are stationary systems that reside under tectonic plates that move over the top of them. A volcanic island can begin to form when magma from the hot spot breaches the ocean floor. However, the rate of plate movement will determine how large the volcanic island is and how long it will be active. A volcanic island that does not breach the ocean surface and is at least 1,000 m tall is called a seamount (Figure 1.7). Although they are not visible at sea level, more than 14,500 seamounts have been identified at the bottom of the ocean, covering an area of almost 9 million km^2. Similarly, a seamount built above sea level

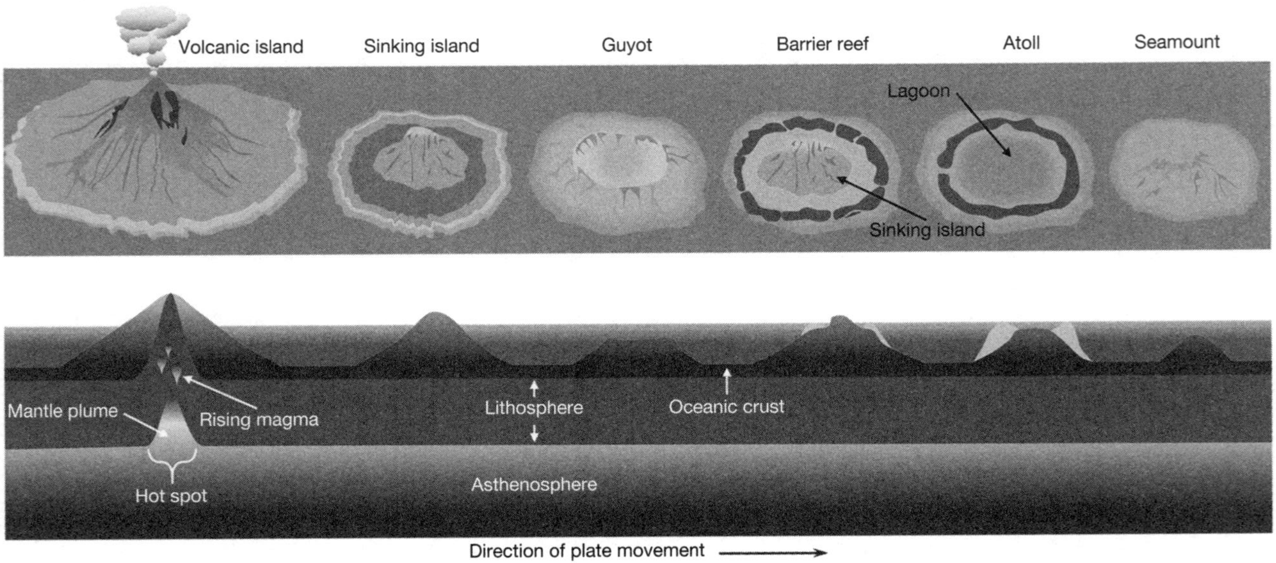

Figure 1.7 Volcanic island evolution as created by hot-spot upwelling. Modified from *Encyclopedia Britannica* (2012).

is called a guyot. Guyots often have flat tops because wave erosion over time destroys the top of the seamount, carving its flattened appearance (Figure 1.7). Seamounts can also form along divergent plate boundaries, like the Mid-Ocean Ridge, although these typically result in effusive sheet flows instead of conical uplifted shapes.

The 6,200-km-long Hawaiian-Emperor seamount chain is one of the world's best-known and intensely studied volcanic island chains (Figure 1.8). Age dating of the oldest seamounts reveals that a hot spot under the Pacific Ocean has been active for the last 70 MY. Around 42 MYA, a drastic shift in plate motion, and maybe slippage in the hot spot itself, caused a 60° bend in the island chain to reposition new seamounts and islands in a southeast orientation. The hot spot is currently under Hawaii ("The Big Island"), which provides

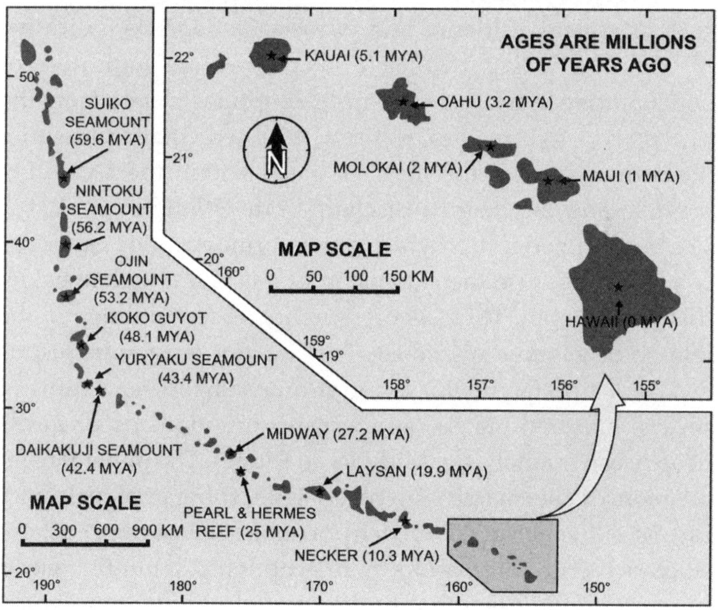

Figure 1.8 Hawaiian-Emperor seamount island chain with correlating age dates. Modified from United States Geological Survey (2020).

ongoing magmatism for Kilauea (1,031 m; 19.4055, −155.2838), Mauna Loa (3,985 m; 19.4730, −155.5911), Mauna Kea (4,149 m; 19.8233, −155.4699), and Hualalai (2,432 m; 19.6905, −155.8652). There is one more active volcano, Mount Haleakala (2,868 m; 20.7205, −156.2552), on the island of Maui. Haleakala is actually the tallest mountain on Earth when measured from the ocean floor to its pinnacle.

Do volcanoes form underwater?

Submarine volcanoes (seamounts) form at the bottom of the ocean by processes similar to those that create land-based versions. Marine geologists estimate that there are 50,000 total seamounts covering the sea floor, which equates to three-quarters of the Earth's volcanic activity. Using remotely operated underwater vehicles (ROVs), geologists can explore submerged landscapes to get glimpses of geomorphic features such as craters, calderas, and pyroclastic deposits indicative of explosive activity. Effusive lava flows are indicative of nonexplosive activity. Submarine eruptions depend on the interplay of four factors: water depth (hydrostatic pressure), the composition of the magma, gases within the ascending magma, and the magma discharge rate (White et al., 2003). The maximum depth at which a submarine eruption can occur is 3,000 m because the expansion of gas is impossible below this water depth. Thus, as pressure increases with depth, the release of gases is restricted. This is why most subaqueous explosions preferentially occur at much shallower depths of several hundred meters, and effusive eruptions occur in the deep ocean (namely, the Mid-Ocean Ridge). The mineral composition of the magma involved with submarine volcanics can be either felsic (silica-rich) or mafic (silica-poor). Those silica-rich eruptions can form an eruption column that ejects hot pumice and gas into the ocean water column. If the explosion is intense (or shallow water depth), a buoyant mixture of pumice, steam, and hot water can reach the surface. Those mafic eruptions typically go undetected at the surface. Yet they can produce thick basaltic magma shaped like pillows at the bottom of the ocean. The overall sum of gases released with submarine volcanics is hard to quantify. However, the last two decades of research on hydrothermal venting have provided some clues as to their source. The gases may be derived directly from the exsolution from subsurface magma or boiling seawater. Comparably, the rate of magma discharged for a particular event is hard to define because it depends on the amount available, which can be linked to plate tectonics or hot spot activity.

Sometimes, a seamount can build tall enough to breach sea level. A seamount 33 km off of the southern coast of Iceland, called Surtsey (119 m; 63.3008, –20.5980), was one of the first documented cases of submarine volcanic development in the world. This eruption took place from 1963 to 1967. It allowed volcanologists to understand the influence of water and magma (termed phreatic eruption), gaining its own formal classification called "Surtseyan-style." This classification has been applied to several other locations where this activity has also occurred (e.g., Taal, Philippines; Kavachi Volcano, Solomon Islands; Ambae, Vanuatu).

Recently, there have been two interesting submarine volcanic events to note. First, a previously submerged volcanic island off Japan, called Niijima (83 m; 34.4377, 134.2413), erupted in October 2023 to create new land above sea level. Next, in December 2021, an eruption began on Hunga Tonga-Hunga Ha'apai (114 m; 20.5638, –175.3808), a submarine volcano in the Tongan archipelago in the South Pacific Ocean. This phreatic eruption emitted an ash plume 58 km tall, generated a tsunami crossing the Pacific Ocean, and produced atmospheric shock waves propagating the entire planet. Volcanologists have compared this mighty volcanic eruption to the 1883 eruption of Krakatoa, Indonesia (463 m; –6.1005,

105.4250), one of the most destructive and deadliest volcanic events in history.

How many active volcanoes are there on the planet today?

By definition, an active volcano is one that has erupted during the last 10,000 years, is currently in an eruptive phase, or has a dynamic subsurface magma chamber that allows it to erupt in the future. A volcano that does not meet these criteria is considered dormant.

Globally, the challenge in quantifying the total number of active volcanoes on the planet is that the definition of a volcano differs as some geologists map individual vents, volcanic edifices, or volcanic fields. However, when a systems approach is applied to the term whereby an eruption is tied to a distinct magma chamber, the number of ~1,350 active volcanoes is agreed upon. About 500 of those 1,350 volcanoes have erupted in historical time. Annually, around 50–70 volcanic eruptions occur with varying intensity and duration. While this is a large number, there is no evidence to suggest that the rate of volcanic activity has increased over time. Instead, the size of populations living near volcanoes and human interest have generated more attention.

About 90% of the world's volcanoes are found along the subduction zone known as the Pacific Ring of Fire (Figure 1.5). We can locate 169 active volcanoes (mostly seamounts) in the United States of America—along the Aleutian Island chain in Alaska, the Cascade Mountain Range in the Pacific Northwest, and the Hawaiian Islands. The last continental eruption in the United States occurred 25 years ago with Mount St. Helens (2,549 m; 46.1911, −122.1944) on May 18, 1980. From 2004 to 2008, Mount St. Helens reawakened, releasing shallow earthquakes and inflating its sides due to magma upwelling. Likewise, there are 82 active volcanoes in Europe, and 33 of these are found on the island of Iceland. Iceland has the highest density of land-based volcanoes and experiences a volcanic eruption every 5–7 years.

The other 10% of active volcanoes are found along other convergent plate boundaries, divergent plate boundaries, or hot spots. There are approximately 50 hot spots in the world today.

INTERPRETING THE LANDSCAPE

VOLCANO GEOMORPHOLOGY PART I

2

WHAT TYPES OF MINERALS AND ROCKS FORM FROM THE SOLIDIFICATION OF MAGMA?

What role does the solidification of magma have in the rock cycle?

The rock cycle describes the transformation of rocks through varying processes that produce three main rock types: igneous, sedimentary, and metamorphic (Figure 2.1). Although cyclical in nature, the rock cycle begins with the origin of magma in the mantle. This buoyant, viscous fluid can make its way up to the surface through fractures in the crust, or it can solidify just below the Earth's surface. The direct solidification of magma to lava forms an igneous rock. Alteration to igneous rocks can occur through varying mechanisms such as weathering and erosion or the application of heat and pressure. When this happens, the igneous rocks are altered in appearance and composition enough to be reclassified as either sedimentary or metamorphic rocks.

Detailed descriptions of igneous rock mineralogy and crystal structure began in the mid-19th century with the invention of the thin section by H. C. Sorby. A "thin section" is a 30-micrometer-thick slice of rock that is attached to a glass microscope slide with glue. A polarized microscope illuminates and magnifies the thin piece of rock so that the mineral structure can be identified (Figure 2.2). The structure and composition of interlocking mineral crystals can provide a petrographer with

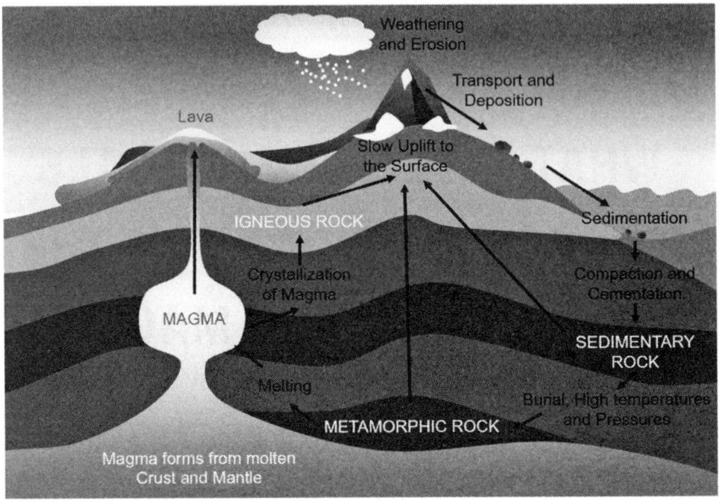

Figure 2.1 Diagram depicting the rock cycle. Modified from Georgia Lou Studios.

Figure 2.2 Photo of a thin section of granite illuminated and magnified by a polarizing microscope. The different colors represent different minerals contained within the granite, namely feldspar, quartz, and biotite. Photo credit: Alamay.

information on how quickly the magma cooled, the types of minerals contained within the specimen, and the depth from which the magma originated.

Magma contains three elements in abundance: oxygen (O^{-2}), silicon (Si^{+4}), and aluminum (Al^{+3}). Therefore, it comes as no surprise that this group of minerals, referred to as silicates, are also the most abundant elements contained within the Earth's crust. As a generalization, the Earth's surface is the result of the solidification of magma into igneous rock forms. Namely, the composition of the Earth's continental crust is granite, and the oceanic crust is basalt.

How do minerals and rocks form from magma?

There are 118 elements found on the periodic table. These elements are dispersed throughout the entire planet or have been found on the surface as remnants of astronomical objects, like meteorites. Magma that originates in the mantle (Figure 1.1) contains elements that will be brought up to the surface by varying processes (e.g., convection cells, hot spots, volcanoes) and will solidify as either a mineral or a rock, depending on its composition. The three most common elements in the mantle are oxygen (44.8%), silicon (21.5%), and magnesium (22.8%). The remaining percentage is primarily made up of iron, aluminum, calcium, sodium, and potassium.

By definition, a mineral is naturally occurring and crystalline and has a specific chemical composition. While lab-created minerals may look like exact replicas, they are not considered to be true minerals because they are synthetic and do not naturally solidify from magma. The crystallinity of a mineral refers to the degree of structural order in a solid whereby the atoms have arranged themselves in a regular and repeating pattern that can best be observed in a thin section (Figure 2.2). Lastly, all minerals have a specific chemical composition. There are two major groups of minerals based on this property: non-silicates and silicates. A non-silicate is any mineral that forms from an individual element, like metals, sulfur, or calcium. Conversely, silicate minerals form from the bonding of silicon and oxygen as $(SiO_4)^{-4}$ to create a foundational building block structure that other elements can adhere to. The silicate group makes up 95% of all minerals within the Earth's crust, including those that are commonplace, like potassium feldspar ($KAlSi_3O_8$), biotite mica ($K(Mg,Fe)_3AlSi_3O_{10}(F,OH)_2$), and the purest silicate, quartz (SiO_2).

A rock is the combination of one or more minerals that originally solidified from a body of magma to form an igneous rock (Figure 2.1). An igneous rock that contains more than 50% silica is referred to as felsic, while those that contain less than 50% silica are mafic (Figure 2.3). Igneous rocks that are felsic are generally lighter in color, tending to be pink, white, or light gray, due to the high silica content. As an example, a common felsic igneous rock is granite. Granite can contain >50% quartz and high amounts of other silica-rich minerals like orthoclase feldspar and plagioclase feldspar. Oppositely, those that are mafic are prone to be black, red, or green due to their higher

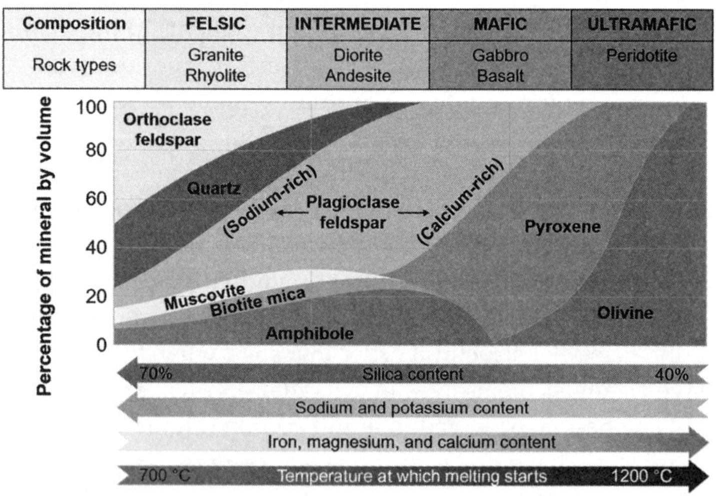

Figure 2.3 Mineralogical content of igneous rocks.

amounts of iron and magnesium. As an example, a common mafic igneous rock is basalt. Basalt can contain <20% quartz but has high amounts of iron-rich minerals like pyroxene, olivine, biotite, and hornblende.

The solidification of minerals that combine to form rocks is described by Bowen's Reaction Series (BRS) (Figure 2.4), named after Norman L. Bowen, a Canadian petrologist who conducted laboratory experiments to determine the melting temperature of common silicate rocks. To understand BRS, imagine that there is a body of magma upwelling from the mantle. This body of magma primarily contains silicon and oxygen but also lesser amounts of all the other elements on the periodic table. When the body of magma reaches a certain depth below the surface (>100 km), the temperature is around 1400°C (Figure 1.3). At this temperature, mafic minerals, like olivine, begin to crystallize out of the body of magma. Likewise, calcium-rich plagioclase feldspar can also begin to solidify. As this body of magma migrates upward toward the surface, it will continue to cool, allowing for other subsequent mafic minerals to form (i.e., pyroxene, amphibole) before those that are felsic (i.e., orthoclase, muscovite). The final mineral to form from this fractional crystallization process is quartz at 800°C.

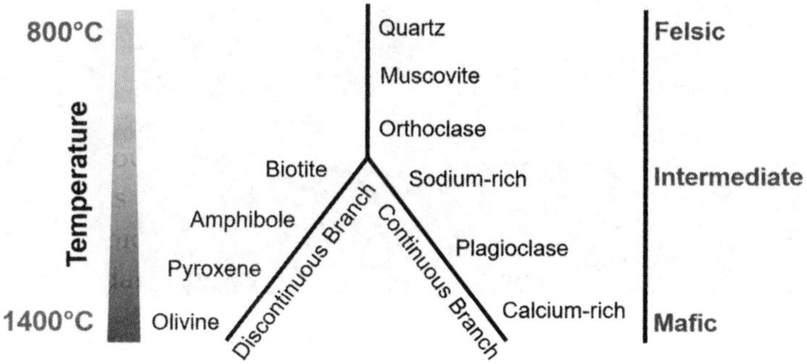

Figure 2.4 Diagram depicting the crystallization of minerals from magma at 1400°C to 800°C. This process is referred to as Bowen's Reaction Series.

The fractional crystallization process can also occur within a volcano's magma chamber (Figure 2.5). Hotter, mafic magmas can begin to form olivine crystals that then settle to the bottom of the magma chamber. The olivine crystals used iron and magnesium in their formation, leaving the remaining magma more felsic. This can stratify the magma chamber, where felsic magma is near the top and more mafic magma is at the bottom. The mixing of these two magmas can form intermediate igneous rocks (Figure 2.3). In a situation where the olivine crystals do not settle to the bottom of the magma chamber, they can be included within cooler magma and brought to the surface by a volcanic eruption. The olivine crystals embedded in a rock matrix are called inclusions and represent a two-stage cooling process.

To summarize, mafic minerals solidify from magma at hotter temperatures than those that are felsic. That is, the magma that contains mafic minerals found in basalt is typically located at greater depths within the Earth. Oppositely,

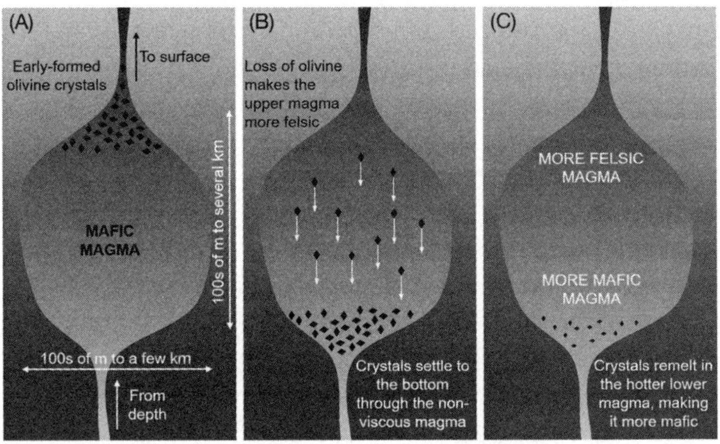

Figure 2.5 An example of crystal settling and formation of a zoned magma chamber. A. Olivine crystals form at top of magma chamber. B. The less buoyant crystals fall to the bottom of the magma chamber. C. The magma chamber becomes fractionated. Modified from S. Earle (2019).

the felsic magma that forms pumice (white volcanic ash) is almost entirely made of quartz and resides at shallower depths in its molten state. Sometimes, due to fractional crystallization, within a single volcanic system, you can have a stratified magma chamber that has cooler felsic magma at the top and hotter mafic magma at the bottom.

What is the difference between magma solidifying at the surface of the Earth or underground?

Igneous rocks are directly associated with the solidification of magma. Magma that upwells from the mantle can solidify in two different locations: within the Earth's crust (termed intrusive) or at the Earth's surface (termed extrusive). Intrusive igneous rocks are also known as plutonic rocks, named for the Roman god of the underworld, Pluto. Likewise, extrusive igneous rocks are referred to as called volcanic, named for Vulcan, the Roman god of fire. The crust of the Earth is made from both processes. As a sweeping generality, geologists often classify the continental crust as intrusive (namely granitic, Figure 2.6A) and the oceanic crust as extrusive (basaltic, Figure 2.6C).

Figure 2.6A Granite. Photographs used for igneous mineral and rock identification. Photo credit: Joann Paracka Marvin.

Figure 2.6B Rhyolite. Photographs used for igneous mineral and rock identification. Photo credit: Joann Paracka Marvin.

Figure 2.6C Basalt. Photographs used for igneous mineral and rock identification. Photo credit: Joann Paracka Marvin.

Often, to form an intrusive igneous rock, there needs to be a preexisting crack, crevasse, or fracture that provides space for magma to grow well. In volcanic settings, it is not uncommon for crustal rifting to occur, which allows for gases or magma to be released at the surface. When these spaces are filled with solid lava, they are referred to as dikes or sills (Figure 2.7).

Figure 2.6D Pumice. Photographs used for igneous mineral and rock identification. Photo credit: Joann Paracka Marvin.

Figure 2.6E Obsidian. Photographs used for igneous mineral and rock identification. Photo credit: Joann Paracka Marvin.

Figure 2.6F Andesite. Photographs used for igneous mineral and rock identification. Photo credit: Joann Paracka Marvin.

Figure 2.6G Diorite. Photographs used for igneous mineral and rock identification. Photo credit: Joann Paracka Marvin.

Figure 2.6H Gabbro. Photographs used for igneous mineral and rock identification. Photo credit: Joann Paracka Marvin.

A dike is a vertical intrusion that crosscuts another rock type. Thus, when looking at dikes in cross section, it can be assumed that they are geologically younger than the rock that surrounds them. Volcanic areas can contain dike swarms hosting several to hundreds of dikes that intrude during a single episode. The world's largest dike swarm (500 km wide, 3,000 km long) is located in western Canada, called the Mackenzie dike swarm (68.4772, −115.9436), and likely resulted from the creation of a

Figure 2.6I Vesicular basalt. Photographs used for igneous mineral and rock identification. Photo credit: Joann Paracka Marvin.

Figure 2.6J Peridotite. Photographs used for igneous mineral and rock identification. Photo credit: Joann Paracka Marvin.

hot spot 1.2 BYA. Similarly, Ernst et al. (2001) describe mafic dike swarms on Venus and Mars.

A sill is a lateral intrusion that is usually fed by a dike (Figure 2.7). While sills are less common than dikes, it is sometimes possible for there to be a horizontal plane of weakness

Figure 2.6K Scoria. Photographs used for igneous mineral and rock identification. Photo credit: Joann Paracka Marvin.

Figure 2.6L Tephra. Photographs used for igneous mineral and rock identification. Photo credit: Joann Paracka Marvin.

Figure 2.6M Pahoehoe. Photographs used for igneous mineral and rock identification. Photo credit: Joann Paracka Marvin.

Figure 2.6N A'a. Photographs used for igneous mineral and rock identification. Photo credit: Joann Paracka Marvin.

Figure 2.6O Sulfur. Photographs used for igneous mineral and rock identification. Photo credit: Joann Paracka Marvin.

between preexisting sedimentary, igneous, or metamorphic rock that erodes, allowing for space for magma to intrude (Figure 2.1). In some instances, sills can be confused with solidified lava flows. However, these systems can be discerned through close inspection of the upper and lower contact surfaces. First, a sill radiates heat in all directions, both contacts

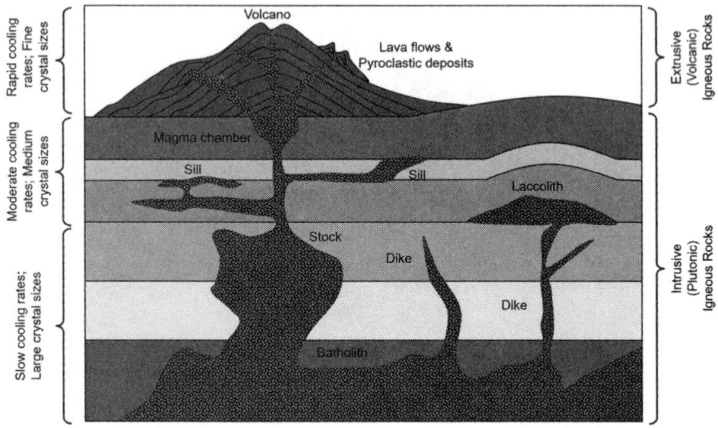

Figure 2.7 Examples of intrusive and extrusive igneous features.

will exhibit partial melting as they have undergone contact metamorphism. Conversely, in a lava flow, only the lower contact experiences melting. Next, lava flows may contain vesicles (Figure 2.6 I), where gases have escaped to the atmosphere. Vesicles are rare in sills because the pressure is capped by the upper contact. Lastly, sills will not show signs of weathering or erosion because they are sheltered from exposure. Other geologic features associated with intrusive magmatism include plutons, laccoliths, and batholiths (Figure 2.7).

An extrusive igneous rock is formed by the cooling of magma on top of the Earth's surface or at the bottom of the ocean (Figure 2.7). There are many different environments that are characterized as extrusive. This includes subduction zones (convergent plate boundaries), divergent plate boundaries, and hotspots.

Subduction zones form when two tectonic plates converge toward each other (Figures 1.5 and 1.6). When an oceanic plate collides with a less dense continental plate, it will begin to melt as the slab is subject to pressure (Figure 1.3), friction, and water. This process produces buoyant magma that can settle in a subsurface magma chamber before it exits onto the Earth's surface during a volcanic eruption. Thus, all features associated with volcanism are considered to be extrusive.

Divergent plate boundaries form when two tectonic plates are moving apart from one another (Figure 1.5). The rifting creates elevated topography on the bottom of the ocean floor, which is referred to as the Mid-Ocean Ridge. The center of the ridge is a low-lying rift valley that acts as a conduit for magma to more easily upwell. For this reason, there is continually brand-new sea floor being produced extrusively at the rift. Another example of magmatic upwelling at the bottom of the seafloor includes black smokers. These hydrothermal vents, rich in rare earth minerals and iron ores, are fissures on the seabed where geothermally heated water discharges.

Lastly, extrusive igneous rocks can be produced at hot spots where magma ponds just below the surface. While tectonic plates are always in motion, a hot spot remains stationary. For this reason, a volcanic chain of islands (Figure 1.7) can be formed extrusively on top of an oceanic plate (Samoa Islands; –13.8611, –171.9558) or the hot spot can support a volcanic region (Yellowstone caldera; 44.4127, –110.7225) if under a continental plate.

How can volcanic igneous rocks be identified?

Within the rock cycle (Figure 2.1), igneous rocks form from the solidification of magma that is either derived from the partial melting of the Earth's crust or the upwelling of mantle material. Igneous rocks are identified by their texture or appearance, which is linked to their mineralogy and rate of solidification. For igneous rocks, the mineralogy is described as either felsic (contains >50% silica), intermediate (50% silica), or mafic (contains <50% silica). The amount of silica present in the rock will influence its color. If a sample is felsic, then its appearance will be light in color, such as white or pink (Figure 2.6 A, Figure B, D). Oppositely, if a sample is mafic, then its appearance will be dark in color: gray, black, red, and green (Figure 2.6 C, E, F, H, I, J).

Samples that are 50% felsic and 50% mafic are called intermediate (Figure 2.6 G). Additionally, there is a classification called ultramafic igneous rocks that commonly form in the mantle and are almost completely made of the mineral olivine and pyroxene (Figure 2.6 J). Next, the location where the magma solidifies also influences its texture. If the magma solidifies intrusively within the Earth's crust, the cooling process is slow, which provides time for mineral crystals to grow in size (>1 mm). Oppositely, if the magma solidifies extrusively at the Earth's surface, the cooling process is quickened as the magma immediately comes into contact with oxygen, allowing little time for mineral crystals to develop (grains <1 mm).

There are five main types of textures used to describe igneous rocks:

1. Phaneritic: a coarse-grained intrusive igneous rock that has mineral crystals >1 mm that are visible to the naked eye. Igneous rocks that have phaneritic texture include granite, diorite, and gabbro (Figure 2.6 A, Figure G, H).
2. Aphanitic: a fine-grained extrusive igneous rock that has mineral crystals <1 mm that are not visible to the naked eye. Igneous rocks that have aphanitic texture include rhyolite and basalt (Figure 2.6 B, C).
3. Glassy: an extrusive phenomenon that occurs when lava is cooled so rapidly that no grains form, resulting in the appearance of glass (Figure 2.6 E). Igneous rocks that have glassy texture include pumice and obsidian (Figure 2.6 D, E).
4. Porphyritic: a texture that results from the combination of intrusive and extrusive cooling. Below ground, crystals will have formed slowly before being included in a magma matrix that has erupted at the surface to cool quickly. The result is an aphanitic rock that contains phenocrysts (crystals) that are within the matrix. An igneous rock that has this texture is porphyritic andesite (Figure 2.6 F).

5. Pyroclastic: texture results from explosive eruptions that eject magma into the air, resulting in small, fragmented pieces. Igneous rocks that have pyroclastic texture include volcanic ash, tephra (Figure 2.6 L), and volcanic bombs.

When identifying igneous rocks, it is important to realize that their appearance can be altered during or after the solidification process. As an example, sometimes gasses are trapped within magma as it is rising to the surface. Upon eruption, the aphanitic matrix will solidify quickly, trapping gas bubbles. The bubbles, or vesicles, give the rock a porous characteristic, called vesicular basalt (Figure 2.6 I). Geologists can estimate the amount of gas released during volcanic events by quantifying the amount and volume of vesicles.

After solidification, all igneous rocks exposed at the surface are subject to physical and chemical weathering, which can alter their appearance. The inverse of Bowen's Reaction Series (Figure 2.4) is called the Goldrich Dissolution Series. The Goldrich Dissolution Series describes the weathering rates of common igneous minerals on the Earth's surface, with mafic minerals that form at higher temperatures being less stable than felsic minerals that form at lower temperatures. Physical and chemical weathering can be especially visible in areas where hydrothermal venting occurs, such as hot springs and fumaroles. The hydrothermal alteration of volcanic host rocks involves the replacement of primary igneous minerals (Figure 2.4) with other minerals that are stable at venting temperatures between 150°C and 400°C. Replacement minerals of quartz or other forms of silica can result in natural beauty like chalcedony, opal, or amorphous silica. Other replacement minerals can provide economically valuable resources like smectite, talc, kaolinite, magnetite, hematite, or goethite. Famous for its hydrothermal alteration landscape, Campi Flegrei, Italy (458 m; 40.8275, 14.1361), has been referred to as the entrance to Hell since ancient history for its orange and yellow landscape.

What are the common rocks and minerals associated with volcanic environments?

Volcanic rocks are primarily named by their mineralogy. The common minerals found in volcanic rocks include quartz, potassium feldspar, plagioclase, muscovite, biotite, amphibole, pyroxene, and olivine (Figure 2.3). Igneous rocks are a composition of minerals. Those found in volcanic environments are separated into three main groups depending on the amount of silica content: rhyolitic (>70% SiO_2), andesitic (50%–60% SiO_2), and basaltic (<45% SiO_2). With that stated, the most common rocks found in volcanic areas include rhyolite, andesite, pumice, obsidian, basalt, and scoria (Figure 2.6).

While rhyolitic and andesitic magmas are straightforward, basalt lavas have variability in their composition due to their origins in different plate tectonic settings. As an example, basalts can be classified as tholeiitic or alkali. Tholeiitic rock types tend to be more enriched in iron and less enriched in aluminum than alkali equivalents. Tholeiitic basalts are sometimes referred to as Mid-Ocean Ridge Basalts (MORBs) because they are associated with ocean spreading centers (divergent plate boundaries) that allow for magma to rise directly from the mantle. This scenario is different from magmas that solidify as Ocean Island Basalts (OIBs) because the magma has to upwell through the preexisting crust, which causes melting and mixing. For this reason, OIBs have higher MgO content and are referred to as alkali olivine basalt. Lastly, collisional zones (convergent plate boundaries) are associated with alkali basalt, which has a higher aluminum content, which results from the melting and mixing of the felsic continental crust (Figure 1.6).

Volcanic systems can produce important minerals that are the result of weathering, epithermal deposition, or submarine or deep mantle volcanic eruptions. An example of a weathering byproduct of basalt is bauxite. Western Australia produces one-third of the world's bauxite, which is used in the production of aluminum.

In areas where warm waters are found at shallow depths (<1 km) and hot temperatures (50°C–200°C), epithermal minerals, like gold, can be deposited. Hot water heated by magma carries dissolved gold toward the surface through fractured rock that will eventually be deposited as a vein. The Ladolam gold deposit was found in 1982 on the volcanic Lihir Island in Papua New Guinea (700 m; −3.1655, 152.5608). The deposit dates to 350,000 to 100,000 years ago and is estimated to contain 42 million ounces of gold.

While many different types of minerals can be found near submarine volcanoes, the most common are lead, zinc, and copper. The bottom of the seafloor creates a unique situation where hot volcanic fluids meet cold seawater, causing the precipitation and leaching of these metals. The deposition of these metals has been observed as they form along the Mid-Ocean Ridge. These "chimney" formations have been sampled, revealing that they contain as much as 29% by weight zinc and 6% by weight copper (John et al., 2008).

An example of a deep mantle volcanic mineral is diamond. Most diamonds form under the high temperature and high pressure setting only found in the asthenosphere between 130 and 200 km below the surface (Figure 1.3), where they reside in a rare ultramafic magma that is a variant of peridotite called kimberlite. At a depth of 200 km, the temperature is 1500°C, and the pressure is 60,000 times greater than at the surface. This zone is referred to as the diamond stability zone. The diamonds are only brought to the surface by deep source volcanic eruptions that form a vent called a diatreme. Although never witnessed by humans, the eruptions are thought to be violent because the kimberlite magma is saturated in carbon dioxide and water. Most diamond deposits date to the Cretaceous Period (146 MYA–66.5 MYA), although they can be found as early as 25 MYA.

INTERPRETING THE LANDSCAPE

VOLCANO GEOMORPHOLOGY PART II

3

WHAT TYPES OF VOLCANOES FORM FROM THE SOLIDIFICATION OF MAGMA?

How hot is lava on the surface, and how quickly does it cool?

In geology, the term "magma" is used for the molten, pliable rock that is stored underground. Once the liquid magma appears on the Earth's surface, it is referred to as lava. The word "lava" was first used to describe the devastation of the AD 79 eruption of Mount Vesuvius, Italy (1,281 m; 40.8230, 24.4288). The word is derived from the Italian word *lavare* (to wash), which was applied to the eruption of Vesuvius as washing away landscapes by burying cities, such as Pompeii.

The temperature of the lava is dependent on its mineral composition. Those lavas that are more mafic will be hotter than those that are felsic (Figure 2.4). For example, the eruption temperature of basalt is between 1050°C and 1200°C, while the eruption temperature of rhyolite is 700°C–900°C (Figure 3.1). Although infrequent, lavas that are sulfur-rich (e.g., Shiretoko-Iosan volcano, Japan, 44.1566, 145.1861, and Lastarria volcano, Chile, –33.4400, –70.6431) and those that are of carbonate composition (e.g., Ol Doinyo Lengai volcano, Tanzania, –2.6600, 35.9133) occur at the surface at lower temperatures (150°C and 600°C, respectively).

The viscosity of the magma is an important variable relating to lava cooling rates. More mafic magmas have a low viscosity or a low resistance to flow. Oppositely, those that are

Composition	Eruption Temperature °C
Basalt	1050–1200
Andesite	950–1170
Rhyolite	700–900
Komatiite	>1600
Water at Earth's surface	20
Temperature °C	**Color of Lava Surface**
1150	White
1090	Golden yellow
900	Orange
700	Bright cherry red
600	Dull red
475	Lowest visible red

Figure 3.1 Eruption temperature of lava as compared to composition and color. Modified from Kilburn (2000).

felsic or silica-rich have high viscosity or are more resistant to flow. Felsic magmas are more restrained. The difference in viscosity will impact how quickly the magma arrives at the surface. It is not uncommon for magma chambers to become differentiated in the subsurface, whereby hotter mafic minerals begin to crystallize out of the magma, allowing the cooler felsic magma to rise (Figure 2.5). Upon eruption, the lava can include phenocrysts that represent formation deep within the mantle. Geologists can age and date the phenocrysts to understand how long the magma was stored below ground before the eruption.

The color of the lava at the surface is a predictor of its temperature; the lighter the color, the hotter the lava (Figure 3.1). The most active parts of flows have a mean surface temperature between 475°C (dull red) and 1100°C (golden yellow). It is essential to recognize that the exterior of a lava flow will cool first, creating a crust. This crust insulates the lava that can still be flowing in a molten state underneath. Once the flow ceases, the time it takes for the piles of lava to cool entirely depends on their thickness. Data from the 2018 Lower East Rift Zone eruption of Kilauea (Hawaii) found that when average flow thickness was 10–15 m, it took roughly 8 months to 1.5 years to solidify, and thicker piles (20–30 m) could take up to 6 years or more (Meredith et al., 2022). This is why it is important not to walk on new lava fields even if they appear to be cooled.

Additionally, lava fields are dangerous to walk on because hollow lava tubes could have formed below the surface during the cooling process, and these could collapse under the weight of a human body. Lava tubes regularly transport basaltic lava flows underground for distances of 10–20 km having an average cooling rate of ≤1°C/km (Kauahikaua et al., 2003). The longest known lava tube in the world is 65 km long and 1,101 m below the surface; it is part of the 500-year-old Kazumura lava tube system in Kilauea, Hawaii (1,222 m; 19.5180, −155.0313).

Why do some volcanoes produce violent eruptions while others do not?

Have you ever noticed that some volcanoes produce tall white ash columns while others produce gently flowing lava? A textbook comparison would be eruptions at Mount Pinatubo, Philippines (1,486 m; 15.1436, 120.3500) versus Kilauea, Hawaii (1,222 m; 19.5180, −155.0313). The difference between these two scenarios is that the first is a violent eruption, and the second is nonviolent.

A volcano's eruption style results from the type of magma that feeds the system, its volatile gas content, and the presence or absence of water. Violent eruptions occur from silica-rich (felsic) magma slowing moving toward the central vent's surface (Figure 1.4). Felsic magma has a high viscosity or high resistance to flow due to the strong covalent bond formed between the silicon and oxygen atoms. The slow upwelling process of magma traps gases (i.e., water vapor, carbon dioxide, sulfur dioxide) that will eventually be released when accumulation reaches a breaking point. If water is present, perhaps held in the caldera, or if the volcano's peak is glaciated, the explosivity is amplified. This is because when the water is heated to a boil or flash point, it will expand its original volume 1,700 times, inducing ignition. Similar to a champagne bottle being uncorked, a violent volcanic eruption can occur immediately, releasing an abundance of energy that can result in a sonic boom, a shock wave, or a tsunami. A nonviolent or quiet eruption occurs when the magma is mafic. Silica-poor magma has a low viscosity and a low resistance to flow, allowing it to be effusively discharged through a vent or fissure to the surface, where it will cool. While mafic magma can still trap gas to later solidify as vesicles (Figure 2.6 I), the volatile content is usually substantially lower than its felsic equivalent. Likewise, when mafic magma intersects water at or below the surface, the energy of the eruption will be intensified (termed phreatic eruption). For the most part, however, effusive eruptions are safe for viewing from a distance and are often called "tourist eruptions."

How are volcanoes classified by their shape (morphology)?

Volcanoes are identified by the shape of their cone, slope of their flanks, and size (height/width ratio). Additionally, a volcano's overall proportions can be linked to its age and eruptive history, with those that are larger usually being older (Figure 3.2). The combination of these variables results in three main volcano classification types: cinder cone, shield, and strato, or composite.

Cinder cones originate when a fissure or vent brings magma to the surface in the form of a fountain (e.g., Wizard Island, Oregon; 2,026 m; 42.9339, −122.1480) or when new vents open on larger, existing volcanoes (e.g., Mount Etna, Italy; 3,270 m; 37.7511, 14.9952). These explosive eruptions contain large quantities of gas that expand, forcing magma vertically into the air, where it solidifies before falling. If the wind is persistently

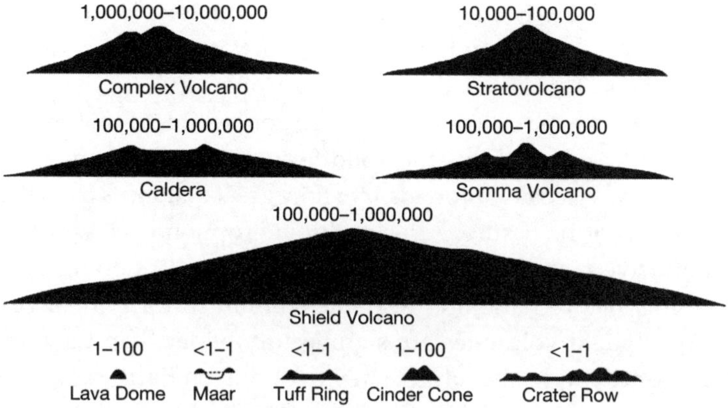

Figure 3.2 Volcano morphology and time line, with relative sizes, shapes, and life spans of various types of volcanoes. The volcano images are vertically exaggerated two times, and the craters and calderas are exaggerated four times. The approximate erosional expectances of the volcanic structures are listed in years. Modified from Decker and Decker (2005).

blowing during the eruption phase, the shape will appear asymmetrical instead of conical. The duration and intensity of the magma production can impact the cone's overall size; half of all historic cinder cone eruptions have lasted 30 days, and 95% have lasted less than a year. A cinder cone volcano can range from tens to hundreds of meters tall and often has a distinct bowl-shaped crater at the top. They are called cinder cones because they are composed of loose cinders, or rock fragments, that solidify quickly when magma explodes into the air. This unconsolidated pyroclastic material can be made of scoria or tephra (Figure 2.6 K and L) that settle around a central crater to build a steep-sided (30°–40°) cone. Cinder cones are monogenetic, meaning that they are the product of a single eruptive event. After the eruption has completed, the vent connecting to the magma source becomes filled with previously made loose cinders and becomes dormant. Cinder cones are different from their counterparts, such as spatter cones, because they are uniform in shape. Spatter cones have a collapsed side where, during fountaining, lava flows down the weaker side, creating a "lava waterfall." Tuff cones, different from cinder and spatter, are formed from the interaction of rising magma with water. This is called a phreatic eruption style.

At hot spots or continental rift zones, shield volcanoes (Figure 3.2) form from the solidification of basaltic magma with a low viscosity, allowing it to flow great distances from the central vent or fissure. After multiple eruption cycles, layers of effusive magma harden to build up a broadly sloping base (2°–10°) before flattening near the summit to form a convex shape. Shield volcanoes are significantly wider than they are tall. For example, the largest shield volcano on Earth is Mauna Loa (4,169 m; 19.4708, −155.5930) in Hawaii. Its base, which is measured on the seafloor, is 160 km wide, which is 20 times the height of its summit. Within these large systems, extensive lava tubes are not uncommon. When lava is exposed to the air, the surface will develop a crust that insulates the flowing magma. The magma follows a path of least resistance, thereby carving a channel. The easy-flowing lava will completely drain the channel once the discharge has dissipated, leaving a hollow cavity or tube. Another hot-spot-generated shield volcano is Piton de la Fournaise (2,632 m; −21.2394, 55.7083) located on the eastern side of La Réunion island in the Indian Ocean. Originating 530,000 years ago, Piton de la Fournaise has experienced three caldera collapses, at 250,000, 65,000, and 5,000 years ago. Today, a prominent feature of this volcano is the 8-km-wide caldera found at the volcano's peak. This volcano has had more than 150 recorded eruptions since the 17th century, with the most recent occurring August 10, 2023.

Stratovolcanoes, also called composite, make up about 60% of individual volcanoes on the planet and are found on convergent plate boundaries (Figure 3.2; Figure 1.5). The name "composite" comes from the volcano's combined deposition of layers that alternate between lava and pyroclastic material. Unlike cinders or shields, stratovolcanoes are primarily associated with felsic magma, which produces andesite (Figure 2.6 F) and dacite. While stratovolcanoes can have multiple vents, the tallest ones have a single eruption site at the summit. Here again, the magma composition dictates the volcano's shape. Silica-rich magma has a high viscosity and a high resistance to flow-trapping gases as it ascends to the surface, causing explosive volcanism. The slow-moving lava cools near the vent, forming the volcano's steep flanks (30°–35°). Most active stratovolcanoes have shallow magma chambers that reside 5–10 km below the ground surface.

Is a caldera a volcano?

A caldera is a volcanic structure that is associated with shield or stratovolcanism. A caldera is a large depression that forms after or during a volcanic eruption (Figure 3.3). When warm, buoyant magma rises to the surface, volcanic structures begin to inflate. Upon eruption, the space formerly hosting the magma empties, causing the volcano's sides to collapse inward. The resulting calderas can range from 1 to 48 km in diameter.

Caldera collapse is both uniform and nonuniform, and therefore, the resulting surface expression has variability, producing five unique classifications that include piston, piecemeal, trapdoor, downsag, and funnel (Figure 3.4). A complete ring representing the volcano's base will form when the volcano walls collapse as a single crustal block. This style, referred to as piston, represents most calderas in North America. Next, piecemeal calderas comprise multiple independent

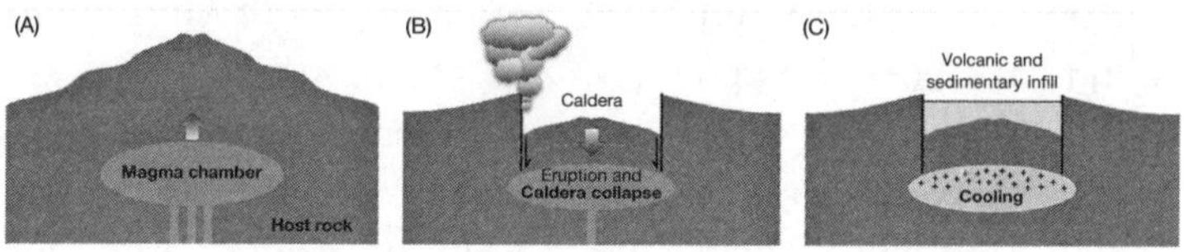

Figure 3.3 Evolution of a caldera. A. Upwelling of magma under volcanic structure prior to eruption. B. The collapse of the volcanic structure post eruption. C. The infilling of the caldera. Modified from Acocella (2021).

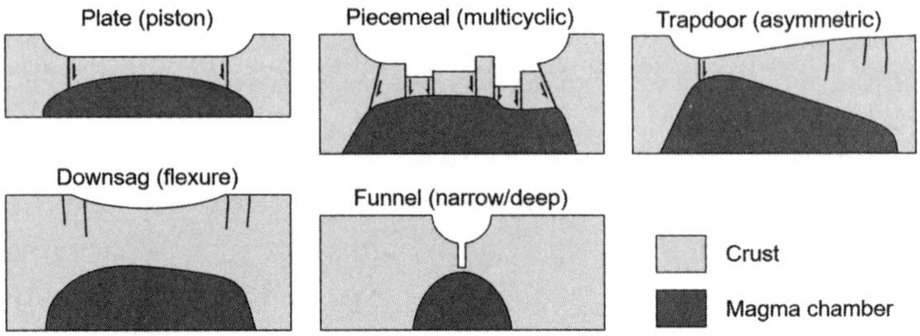

Figure 3.4 Five different types of calderas. Modified from Acocella (2021).

blocks of rock that have collapsed along the floor of the caldera when subsidence becomes uneven. It is not uncommon for slipping blocks to generate low-level seismicity that can initiate landslide activity along the caldera walls. Additionally, the blocks provide conduits for excess heat to be released at the surface, which makes the surrounding environment susceptible to hydrothermal alteration. Trapdoor calderas are best identified by their distinct asymmetrical appearance, allowing them to be confused with craters made by meteorites. In this scenario, the incompleteness of subsidence causes one side of the caldera to be at a higher elevation than the opposing side. Downsag calderas occur when the magma chamber is underground (>10 km) and has less volume. While the area above the magma chamber does subside, a ring fault does not form at all, at least initially. Later, compaction over time can cause the basin to sink, creating a newly formed ring. Lastly, funnel calderas are classified by a "V" or funnel-shaped depression. A funnel caldera can form from the collapse of a piecemeal caldera, as was observed in the 2000 eruption of Miyakejima Island, Japan (815 m; 34.0794, 139.5175) (Nobuo, 2008).

In some situations, uplift of the caldera floor can occur as magma upwelling refills the emptied magma chamber following subsidence. The result is a resurgent caldera. Although it has a dome-shaped appearance, it differs from a lava dome because it is an uplift feature. Resurgent calderas, sometimes

Volcano Name	Latitude and Longitude	Eruption Year	Volcanic Explosivity Index	Notes
Hunga Tonga-Hunga Ha'apai, Polynesia	20.5638, -175.3808	2022	5.7	Sonic boom heard in Alaska, shock waves circled planet twice
Krakatoa, Indonesia	-6.1005, 105.4250	1883	6.0	Tsunami wave (40 m tall) killed 36,000 people
Santa Maria, Guatemala	14.7580, -91.5594	1902	6.0	Largest eruption in 20th century
Novarupta, Alaska	58.2283, -155.1230	1912	6.0	160 km away, the sky was black with ash for 60 hours
Ambrym Island, Vanuatu	-16.2311, 168.1555	A.D. 50	6.0+	Formed a 12 km wide caldera
Ilopango, El Salvador	13.7092, -89.1068	A.D. 431	6.0+	May have initiated movement of Mayan culture from highland to lowland
Mount Thera, Greece	36.4072, 25.3952	1610 B.C.	7.0	5 times larger than Krakatoa and created massive holes in Santorini
Tambora, Indonesia	-8.2188, 117.9938	1815	7.0	Eruption heard 1,930 km away; pyroclastic flow killed 11,000
Toba, Indonesia	2.837, 98.6261	74,000	8.0	Lead to severe global cooling
Yellowstone caldera, WY, USA	44.4125, -110.7241	640,000	8.0	Created crater 48 by 72 km

Figure 3.5 Top 10 largest volcanic episodes.

referred to as "supervolcanoes," are among the most extensive volcanic features on Earth because they can span tens of kilometers in width. Some of the most cataclysmic volcanic episodes are associated with resurgent calderas, including Yellowstone (Wyoming) 640,000 years ago (2,400 m; 44.4125, –110.7241), Toba (Indonesia) 74,000 years ago (905 m; 98.6261), and Krakatoa (Indonesia) in 1883 (463 m; –6.1005, 105.4250) (Figure 3.5). A smaller and lesser-known resurgent caldera is the Long Valley caldera in California (2,300 m; 37.7177, –118.8822).

What is a lava dome?

Lava domes are conical mounds of felsic magma (andesitic or rhyolitic) that solidify near the vent of a volcano or along a plane of weakness in the volcano's structure, like a flank or fault line (Figure 1.4). Lava domes are usually created near the end of a volcanic episode, when the magma has lost most of its gases, leaving the viscous magma with little explosive energy to move it great distances before it crystallizes. Lava domes have unpredictable patterns of growth that permit them to accumulate in size from a few meters to several kilometers over months, years, or even centuries (Calder et al., 2015). As an example, over the course of 4 years (2004–2008) a lava dome was built at Mount St. Helens, Washington (2,367 m; 46.1888, –122.1975), amounting to more than 44 million cubic meters.

These structures are classified into four main categories by their morphology (shape), surface texture, and eruptive style: torta, peléean, coulées, and upheaved plug. The lava dome referred to as torta is found on low-lying or mostly flat ground surfaces that are common in the Andes Mountains of South America. Chillahuita, Chile (4,706 m; –22.1694, –68.0416), provides a textbook example of a torta lava dome. As the lava rises to the surface within the dome's center, it will push away older material, forming an onion-like internal structure. Next, peléean lava domes are known for their steep-sided appearance. They are circular like torta-style domes, but instead of having a flat top, they have a smooth upper surface that is punctuated by tall vertical spines. The vertical spines are often unstable and crumble to create talus slopes. Peléean lava domes are associated with large stratovolcanoes and were closely documented forming at Mount St. Helens, Washington, from 2004 to 2006. Coulées are found on steep slopes and are a blend between a lava flow and a dome structure. Upon eruption, the highly viscous magma remains somewhat stationary and typically does not flow more than a few kilometers downslope. An example of a coulée is the Cerro Chao, Chile (400 m; –22.1194, –68.1561). Lastly, the dome referred to as an upheaved plug is rare. These occur when immobile magma is forced to the surface like a piston, often carrying the country rock with it. An example is Shizan, Japan (398 m; 42.5425, 140.8644).

While seemingly docile during their growth period, lava domes typically form during volcanoes' eruptive cycle, which can be a precursor to an upcoming explosive phase. The sudden release of gases during an explosion can cause the collapse of a lava dome. This force initiates a surge of hot gases intermixed with previously deposited ash that becomes mobilized, creating a pyroclastic flow. And pyroclastic flows from lava dome collapse caused the most volcanic-related fatalities in the 20th century (Fink and Anderson, 2023).

Why do some volcanoes have a crater lake? Furthermore, why are some lakes pristine while others are acidic? Why do some craters hold lava?

In some climatic settings, a volcano's caldera (or crater) can act as a catchment basin for water. Often, the basin becomes impermeable due to the accumulation of pyroclastic material. This creates a barrier at the bottom of the caldera, allowing it to hold water via rainfall, snow melt, or groundwater. However, the quality of the water held within craters can vary.

The deepest (592 m), and perhaps most pristine, crater lake in the United States is on top of Mount Mazama (3,700 m; 42.9402, −122.1436), a stratovolcano within the Cascade Mountain Range in Oregon. This crater, referred to as Crater Lake, was formed by a cataclysmic eruption 7,700 years ago estimated to be 119 times larger than the 1980 Mount St. Helens eruption. Regarding hydrology, the system is defined as a closed basin lake because it does not have an inlet or outlet to a stream. Thus, like many other lakes held within a caldera, Crater Lake is incredibly clear because it does not contain suspended sediment within its water column.

Other crater lakes, like Lagunas Caliente at Poás Volcano (2,708 m; 0.1975, −84.2308) in Costa Rica, can have variability in color (turquoise, yellow) due to their extreme acidity (pH 1.0–1.5). This is the consequence of its shallow magma chamber continuing to be filled with sulfur-rich magma, which heats the lake water to between 38°C and 90°C. Under these heated conditions, volcanic gases (hydrochloric acid, hydrogen sulfide, sulfur dioxide) will exsolve from the magma as it rises through the well-developed hydrothermal system to interact with the lake's basin water, causing the water to become turquoise to yellow. Before a volcanic eruption at Poas, the lake water will turn gray as the sediment at the bottom gets mixed with magma upwelling. Thus, the color of the water is often used as a predictive tool. However, only 600 m down-gradient of Lagunas Caliente there is an inactive caldera at Poás (Laguna Botos) that maintains a temperature of 23°C–28°C with a typical acidity (pH 7.0–8.0), yet it is clear in color as a result of its healthy population of at least 23 species of planktonic algae.

Sometimes, a crater can even hold lava! Famously, the deep (793 m) summit basin of Mount Nyiragongo, Democratic Republic of the Congo, Africa (3,470 m; −1.5220, 29.2505), hosted a lava lake for half a century. However, on January 10, 1977, a 6-km-long fissure opened up on the southern flank of this stratovolcano, allowing the lava lake to suddenly empty. Another eruption in 2002 refilled the lava lake, which persists today. Although magma discharge rates increase and decrease over time, the lava lake at Mount Nyiragongo has the largest recorded volumetric supply in the world.

Do rootless cones (pseudo craters) form instantaneously?

Yes, rootless cones form instantaneously when hot lava comes into contact with a water-saturated surface, such as a wetland, lake, river, or coastal environment. The lava traps the water (or ice) in the underlying sediment until it reaches boiling. This causes a one-time steam explosion that leaves a crater that is very similar in appearance to a cinder cone in size (1 to 100 m diameter), shape (conical), and composition (ash, scoria) (Figure 3.2).

A cinder cone is an independent structure, but a rootless cone is commonly part of a complex of many cones that outline the pathway of the effusive lava flowing across an area from high elevation to low. The number of rootless cones and the distance to the nearest neighbor are limited by the amount of water present at each site and the ease of steam to be released at the surface.

While rootless cones are widespread in Hawaii, the Galapagos Islands, and throughout the Columbia River Basalts (Oregon, Washington, Idaho), Iceland has the most rootless cones on Earth, with more than 13,000 cones created by six eruptions (Boreham et al., 2018). Of these, the Skútustadagigar (65.5708, −17.0341) rootless cone complex near Lake Myvtan is the textbook example.

What is a maar?

Another type of crater, called a maar (Figure 3.2), forms when magma uprising toward the surface through a crack (diatreme) intersects groundwater to cause a shallow explosive eruption. The eruption leaves a surficial depression ranging from 20 to 3,000 m in diameter and 5 to 200 m in depth. The craters are often surrounded by a ring of ejected ash and angular, nonvesicular, pebble-sized tephra material, although this material can erode over time. While cinder cones and rootless cones are commonly made of basaltic material, maars can be formed of other magma types, including carbonatitic, intermediate, or andesitic (Figure 2.3). Another difference is that maars often fill with water to form a lake. The level of the lake represents the level of the groundwater table.

A deep understanding of how groundwater maars form came from the observed 10-day phreatic eruption in April 1977 that created two craters in Ukinrek, Alaska (57.8319, −156.5111); the largest crater is 300 m wide and 70 m deep.

Can fissure eruptions happen anywhere?

About 8% of volcanic activity in the world is classified as a fissure eruption (Newhall and Self, 1982). Fissure eruptions occur when linear underground passageways, called dykes, get infilled with magma. These volcanic systems can be associated with a central volcano, or they can be fed directly from a magma reservoir (Figure 3.6). It is not uncommon for fissure

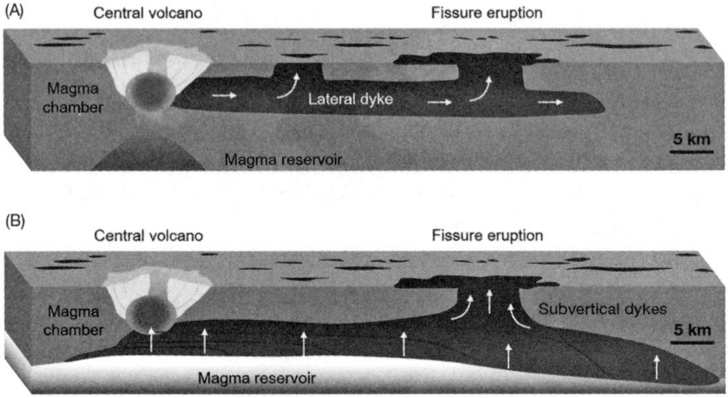

Figure 3.6 Schematic diagram illustrating two scenarios of magma generation for fissure eruptions. A. Linked to the magma chamber of a central volcano. B. Linked to mantle magma reservoir. Modified from Hartley and Thordarson (2013).

eruptions to be associated with hot spots and rifting activity. An area likely to have a fissure eruption is called a fissure swarm and is given a name for reference.

While fissure eruptions cannot be predicted, there are two essential variables that act as precursors to a fissure eruption: ground inflation and significant earthquake activity. When buoyant magma rises toward the surface, the area will lift several centimeters to several meters. Concurrently, the magma is looking for the path of least resistance during this upwell period. As it moves through cracks in the crust, it will release energy, causing low-level seismic activity ($M < 4.0$) at shallow depths (1–4 km below the ground surface). More than 10,000 earthquakes can be recorded prior to the magma penetrating the surface to create a large fissure (crack), allowing low-viscosity magma to breach the surface. The fissures are variable in length and can become longer throughout the eruption. Due to their linearity, early Icelandic Vikings would refer to these eruptions in their sagas as "curtains of fire."

Fissure eruptions do not typically form a central cone but do form vents where magma is centrally concentrated. These vents are also referred to as spatter cones and are formed from pieces of lava cooling in the air and falling to the surface as tephra (Figure 2.6 L). When fissure eruptions occur in real time, geologists monitor two parameters: fountain height and discharge. The fountain height is how high the lava is being forced into the air, and the discharge is how quickly the lava flows on the surface.

Fissure eruptions are often called "tourist eruptions" because the mafic, low-viscosity lava gently follows the topography gradient, making them somewhat predictable. These eruptions are often safe to view from a distance if gases are carefully monitored by experts who are qualified to do so.

How long does an eruption need to last to create a basalt flat/plateau?

Fissure eruptions can vary in intensity and duration depending on the magma held within the reservoir (Figure 3.6). Those that are associated with a hot spot theoretically have an unlimited magma source; however, the time it takes to infill the subsurface dykes is specific to each episode and depends on the magma's viscosity, presence or absence of water and gases, and depth of origin below the surface. The temperature at which minerals crystallize out of a body of magma, as identified by Bowen's Reaction Series (Figure 2.4), can provide helpful clues. As an example, sometimes the magma that is brought to the surface will include olivine or calcium-rich plagioclase crystals that form within the dyke or magma chamber (Figure 2.5). These crystals can be isotopically age-dated, providing an indication of how long the magma has been stored before the eruption.

Over geologic time, there have been several significant fissure eruptions worldwide. These events can last a million years or more and have brought enormous volumes (100–1,000+ km^3) of mafic magma continuously to the surface, "flooding" an area with basalt (Figure 2.6C). While poorly understood, the large quantities of magma generation are attributed to initial hot-spot development supported by mantle material upwelling.

Flood basalt deposits are often called "traps," a word derived from the Swedish language meaning staircase." Because flood basalts pile on top of themselves, they can have a staircase appearance, with the oldest layer on the bottom and younger layers progressively moving toward the top. Each flow can build up to 1,000+ m thick and will solidify as an independent unit to provide a distinct lateral organization that allows the flows to be counted if viewed in cross section.

The three world-class examples of flood basalt deposits include the Siberian Traps (Putorana Plateau, Russia; 68.7255, 96.1300), Deccan Traps (Western Ghats, India; 10.1661, 77.0061), and Columbia River Basalts (WA, OR, ID, United States; 45.7088, −121.7530). The flood basalt event associated with the Siberian Traps occurred 250 MYA and is often linked as a cause for the mass extinction found at the Permian-Triassic boundary. Larger than the Deccan and Columbia episodes combined, the Siberian Traps released a lava flow that covered 7 million km^2 with a total unit thickness of 3,600 m. The event lasted 2 million years and emitted significant amounts of greenhouse gases into the atmosphere, forcing global warming and ocean acidification.

The Deccan Traps formed approximately 66 MYA, at the end of the Cretaceous Period. It is estimated that this event produced a lava flow covering half of India. However, two-thirds of this flow was destroyed during the last Pangea cycle, leaving behind an area of 500,000 km^2. Although this flow lasted only 30,000 years, the total thickness is over 2,000 m, indicating a high discharge rate during its formation. Correlating to the mass extinction of the dinosaurs, the gases released during the eruption may have exacerbated an already changing climate initiated by meteorite impact.

Connected to the Yellowstone Hot Spot, the most recent flood basalts were deposited in the Pacific Northwest region of the United States during the Miocene and early Pliocene, 17.5–6 MYA. The Columbia River Basalts expanded over 210,000 km^2 with an average thickness of 2,000 m, but up to 3 km. With the additional weight of the lava at the surface, parts of the landscape began to subside, creating the Columbia Basin. This low-lying area acts as a catchment for the Columbia River.

INTERPRETING THE LANDSCAPE

VOLCANO GEOMORPHOLOGY PART III

4

WHAT VOLCANIC FEATURES RESULT FROM THE SOLIDIFICATION (OR INFLUENCE) OF MAGMA?

How do hot springs, geysers, mud pots, and fumaroles form?

All of the systems above are directly linked to the presence of water stored underground in an aquifer. When precipitation falls to the surface, it will infiltrate the soil or rock (aquifer), eventually recharging the groundwater table. The rate of recharge of the groundwater table largely depends on the amount of rainfall or snowmelt entering the system, along with the porosity and permeability of the aquifer. The porosity of an aquifer refers to the amount of water stored within a system, and the permeability relates to the ease of removal. Volcanic systems composed of vesicular basalt (Figure 2.6 I) can be highly porous, making them ideal aquifers that can hold large quantities of water and recharge quickly.

When there is a shallow supply of magma, the water in the aquifer can become heated, resulting in hot springs, geysers, mud pots, and fumaroles, which are indicative of active volcanism (Figure 4.1). First, hot springs (water temperature >21°C) are created when subsurface water is geothermally heated. Following the landscape's topography, this hot water

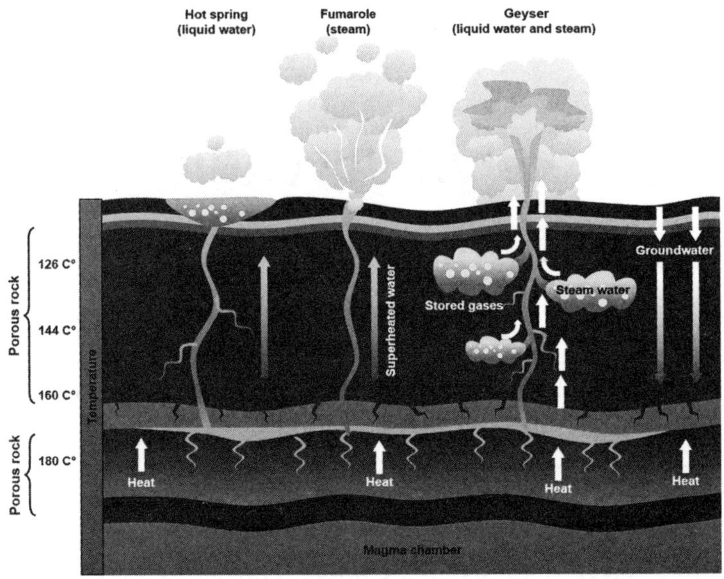

Figure 4.1 The distribution of subsurface water to create hot springs, fumaroles, and geysers. Modified from VectorMine-Dreamstime.com.

can infiltrate springs, rivers, or depressions that act as hot tubs. While these natural features are often enjoyable, it is best to take caution when approaching hot springs because the water can become superheated to temperatures that can cause injury or death upon contact.

Although unusual, areas that have hot springs can have intermediate water spouts called geysers. The word *geyser* was first used in 1597 by an Icelandic farmer describing water shooting from the ground. The word *geyser* can be translated to mean "gusher" or "the one who rages" and is related to the Icelandic verb "gjósa," which means to erupt. A geyser depends on specific hydrologic settings that allow a body of water to be stored underground in a cavity. When this subsurface pocket of water heats to boiling temperature, it will flash to steam and erupt as a water spout at that surface. The water will fall to the ground, funnel down the throat of the geyser, and refill the underground cavity, allowing the process to repeat itself. The time it takes for the water to reach the boiling point is also the time that passes between eruptions. The famous Yellowstone Geyser (Yellowstone National Park, WY; 44.4605, –110.8277) erupts up to 8,400 gallons of boiling water up to 55 m high approximately 20 times daily.

In geothermal areas where water supply has become reduced, mudpots and fumaroles can form. The limited water supply becomes acidic due to hydrogen sulfide (H_2S), allowing it to decay the surrounding rock into clay. The gas escaping at the surface causes the mud to appear as if it is boiling. Where there are mudpots, there are usually fumaroles. Fumaroles are cracks in the Earth's crust that allow steam and other gases, like H_2S, to be emitted into the atmosphere. If hydrogen sulfide is emitted into the air, there will be an unpleasant smell of rotten eggs. This smell is welcome around volcanoes because it means that the sulfur dioxide (SO_2) gases rising out of the magma have had time to be slowly filtered through the groundwater table, meaning that an eruption is not immediately pending. Although bothersome, H_2S is not dangerous to inhale, unlike its counterpart, SO_2. Sometimes, the mudpots or fumaroles will be outlined with precipitating yellow sulfur (Figure 2.6O).

Historically, these natural features have been necessary to human settlement. Some of the first geothermal baths were founded in AD 70 by Romans in Bath, England (51.3808, –2.3600). In Iceland, a 12th-century structure still stands that represents a geothermally heated bathhouse used by a famous saga writer, Snorri Sturluson (1148–1241), who lived in Reyholt (64.6638, –21.2911). Likewise, Europe also experienced a boom in using geothermally heated water for leisure in the 16th century. For example, the city of Budapest, Hungary, is nicknamed the thermal capital of the world, as there are 123 hot springs, representing only 10% of the total area found within the country.

What are the geologic features associated with mafic volcanism?

Mafic magmas have lesser amounts of silica (<50%) and higher amounts of magnesium and iron. This silica-poor mineralogy allows for the magma to have a low viscosity or low resistance to flow. Due to its effusive capabilities, once the magma reaches the surface, it can move fluidly, following the topography of the landscape.

Basaltic (Figure 2.6 C) lava flows generally pass on the surface as broadsheets or channels or travel through lava tubes. Lava that is associated with sheet flows is linked to eruptions that have high discharge rates over low-slope terrain. With little impediment, sheet flows can extend for several to tens of miles. The rates of discharge and volume can decrease throughout an eruption cycle. When this occurs, it is not uncommon for a lava river to form on top of the preexisting sheet flow that will cause melting and form a distinct new channel. Similarly, the lava, looking for the path of least resistance, may follow the outline of a precursory riverbed. Both sheet and channel flows end in delta-like flow fronts.

Lava tubes can form within either of these flow styles. Sometimes, as a sheet flow begins to cool at the surface, a lava river can find a point of weakness underneath the newly formed solid crust (Figure 3.1). The movement of the lava river can erode a linear cavity. Once the discharge has stopped, the empty cave remains in place. Lava tubes that form within a channel are a bit more complicated. The quickly moving lava river can build distinct walls or levees. Eventually, the channel will get deep enough to form an upper crust that will insulate the molten lava, creating a conduit for the flowing lava until the eruption ends. If the flow dissipates slowly enough, it can form flow ledges or lines within the lava tube that will be preserved on the cave walls.

Like limestone caves, lava tubes can also contain speleothems that form from the dripping of molten lava. Lava tubes range in width up to 15 m and can be up to 15 m below the ground surface. While some lava tubes collapse before being preserved, some have long, continuous open passageways that extend for several kilometers and can be visited (Lava Beds National Monument, Tulelake, California; 41.7747, −121.5063).

There are three prominent crustal appearances of basaltic magma seen in the lava flows mentioned above: pahoehoe, a'a, and blocky. The term "blocky" is used to describe a transition between the first two forms (Figure 4.2). "Pahoehoe" and "a'a" are Hawaiian terms introduced in the 19th century to describe common lava types found on Mauna Loa (4,169 m; 19.4719, −155.5936) and Kilauea (1,222 m; 19.5180, −155.0313). Pahoehoe

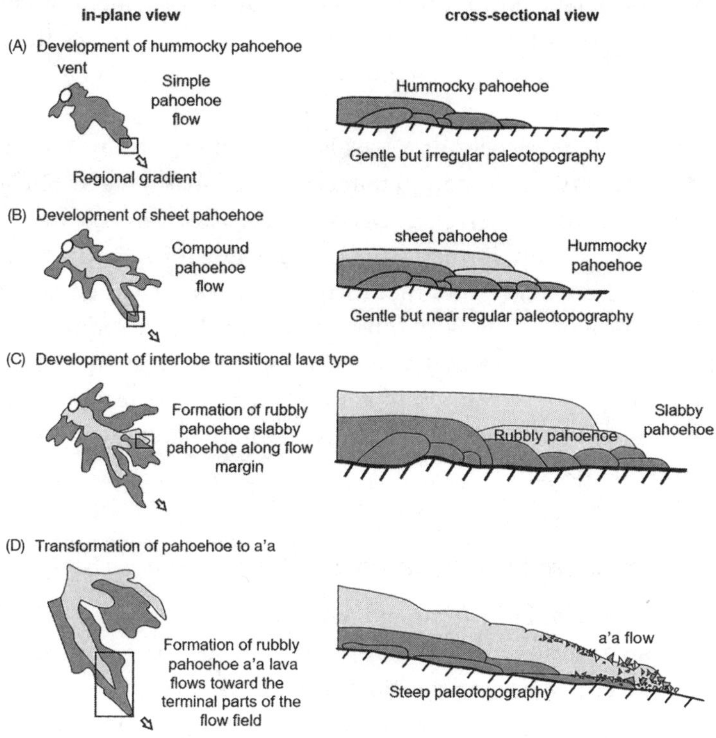

Figure 4.2 Illustration depicting in-plane and cross-sectional view of pahoehoe structures and their transitional state to a'a. Modified from Duraiswami et al. (2014).

lava flows create wrinkles and twists that mimic intertwined pieces of rope (Figure 2.6 M; Figure 4.2). Because of its low velocity, the lava is cooled by the air; this forms a thin crust on the surface of the advancing flow that creates individual semicircular lobes. A'a lavas (Figure 2.6 N; Figure 4.2) are associated with higher discharge rates and steeper slopes than pahoehoe. Because this lava moves faster, the cooled surface becomes irregular and jagged. The rough clasts of lava are referred to as clinkers. Clinkers may fall in front of an advancing flow that will then bury them as it moves, leading to rubbly zones at the top and bottom of the flows (Figure 4.2).

Moreover, a'a is what you might say if you were walking on this type of lava flow with bare feet! It is also worth mentioning that the color of the pahoehoe and a'a structures reflects their relative age. Those that are older will appear reddish because the iron within the lava has oxidized.

If you look closely at a basaltic lava flow, you might be able to identify a formation that is referred to as Pele's Hair, a name that is derived from the Hawaiian goddess of volcanoes, Pele. Pele's Hair is made up of delicate, thin (0.001 mm) shards of volcanic glass that form when bubbles of gas near the surface of a lava flow burst to stretch the skin of molten lava into long threads that may be up to a meter in length. While they can be found immediately after an eruption, their delicate nature makes them easily erodible when exposed to the elements.

What are the geologic features associated with felsic volcanism?

Volcanoes that eject felsic magma are characterized by their explosive nature due to the high viscosity of the magma that forcefully rises to the surface. Once the magma exits the volcano's vent, it will cool and then solidify in the atmosphere as tiny particles of silica termed tephra (Figure 2.6 L). The smallest pieces of tephra are sometimes referred to as volcanic ash (Figure 1.4) and can be less than 0.001 mm in diameter, allowing for these small and light particles to be carried easily long distances by air currents as eruption clouds (Figure 1.4). It is not uncommon for explosive eruptions, like those described in Figure 3.5, to have an eruption column (Figure 1.4) several kilometers tall. As an example, satellite imagery from the Hunga Tonga-Hunga Ha'apai eruption in January 2022 confirmed that the column was the highest recorded, at 57 km above sea level. This was the first recorded eruption to have escaped the stratosphere and entered into the mesosphere layer of the atmosphere, recalling that the troposphere is sea level to 10 km, the stratosphere is 10–30 km, and the mesosphere is 30–60 km. An eruption that produces volcanic ash can also produce larger tephra (pumice) fragments that range in size from less than 2 mm (Figure 2.6 L) to more than 1 m in diameter (called volcanic bombs). Although vesicular, these clasts have more density due to their larger size so they often settle nearest the vent and aid in the formation of the steep flanks attributed to stratovolcanoes (30°–35°) (Figure 3.2).

Sometimes, in felsic eruptions of stratovolcanoes, a pyroclastic flow (Figure 1.4) can form. By definition, a pyroclastic flow is a dense, destructive mass of hot ash, tephra, and gases ejected explosively from a volcano that advances downgradient at velocities up to 50 km/h. Flows are initiated by three main mechanisms:

1. If the volcano does not have enough energy to eject the ash and tephra high into the atmosphere, the eruption column can collapse. This causes the debris to rapidly collect near the steep-sided vent, where it accumulates before gravity forces downward movement.
2. If the volcanic eruption is wholly depleted of gases, all of the ash and tephra will collect near the vent. This process is called "boiling over." Similar to the first scenario described, the material will pile at the vent to an angle of repose before mass movement will occur.

3. A lava dome can form on the flank of a stratovolcano when localized magma accumulates, creating a bulge. If oversteepening of the lava dome occurs or pulses of gases are emitted at the site, a pyroclastic flow can be instigated.

Once the flow is underway, it will follow the topographical path of least resistance, which is commonly down a valley. Most pyroclastic flows will consist of two distinct layers. The basal, or ground flow, includes varying class sizes of tephra from ash to boulders. At the same time, the upper layer is a turbulent cloud of fine-grained ash. Both systems have extreme force that can knock down, bury, or carry away all objects within its pathway. Additionally, the rocks and gases inside the flows are at temperatures between 200°C and 700°C. Pyroclastic flows will continue until their energy dissipates on low-lying topography. This can mean that a flow deposit can be several to hundreds of kilometers away from the vent. After the movement subsides, the unconsolidated material within the flow can be welded together through settling and compaction processes to form an ignimbrite deposit. The degree of welding depends on the weight of the overlying material, so usually within a deposit the upper portion will remain vesicular. At the same time, the bottom becomes less and less permeable. The final deposit can be anywhere from 1 m to 200 m thick. New Zealand's most active volcano, Whakaari, or "White Island," experienced an eruption and launched a pyroclastic flow on December 9, 2019, that resulted in 21 fatalities.

What features are formed when you have a volcanic event under water or ice?

The presence of water in either liquid or solid form can drastically change the solidification rate of lava and, therefore, will influence the type of structure that results. Note that the results differ from those of phreatic eruptions, which are steam-driven explosions that occur when water beneath the ground or on the surface is heated by contact with magma, lava, hot rocks, or new volcanic depositions (e.g., tephra and pyroclastic flow deposits). Instead, these geologic formations are the consequence of low-viscosity mafic magma that comes into contact with water in the ocean, lake bed, river, or glacial ice as it rises to the surface. Thus, structures that remain in situ help geologists interpret past environments and provide valuable clues as to how climate has changed over time. Additionally, isotopic dating techniques of lava can assign age to the landforms, allowing further interpretation of events such as sea-level rise or glaciation.

Two main types of structures occur when mafic lava comes into contact with water: pillow basalts and columnar basalts. Pillow basalts (Figure 4.3), which are named after their inflated appearance, develop when lava erupts underwater, namely at Mid-Ocean Ridge or hot spot localities. The water quickly cools the lava's exterior, forming a crust that insulates the internal magma that is still ascending to the seafloor. The magma

Figure 4.3 Photograph of pillow basalts in Iceland. Photo credit: Alamay.

inside the "pillows" will eventually bust through the hardened exterior, and the process will continue to form interconnected lobes. Pillow basalts are discerned by the vesicles in the structure being found toward its upper surface, which has a convex shape with a narrowing base. Metamorphosed pillow basalts found in Greenland (64.1400, −50.1560) have been age-dated to 3.8 BYO, indicating the presence of significant water bodies on the planet during the Archean Eon. Likewise, columnar basalts also indicate the presence of water during the lava solidification process. When magma rises toward the surface and comes into contact with water, the rate of lava cooling is expedited, and it begins to contract at centers that are equally spaced, allowing a hexagonal fracture pattern to develop that is similar to the formation of mud cracks (Figure 4.4). The fracture pattern that forms at the cooling surface will tend to be propagated down the lava as it cools, forming long, geometric columns perpendicular to the surface flow of water (on top of the lava) that can range in size within a deposit. A shift in colonnade-type columns (e.g., those columns that are straight) to entablature (e.g., those columns that are irregular) may be due to changes in water flow direction or intensity, whereby flow changes create different angles of cooling fronts (Figure 4.5). Thus, the division of colonnade and entablature is the

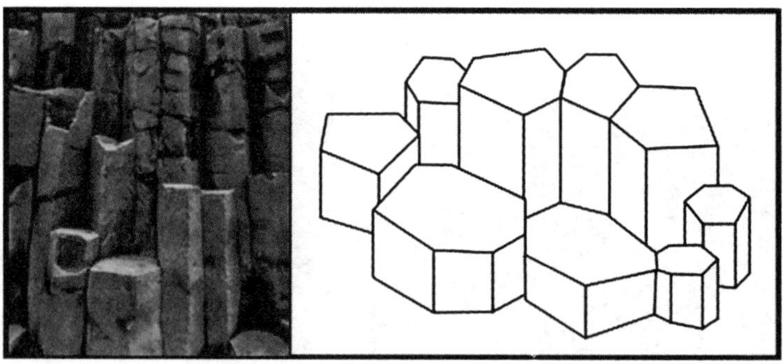

Figure 4.4 Hexagonal columnar basalt. Modified from Jovanelly (2020).

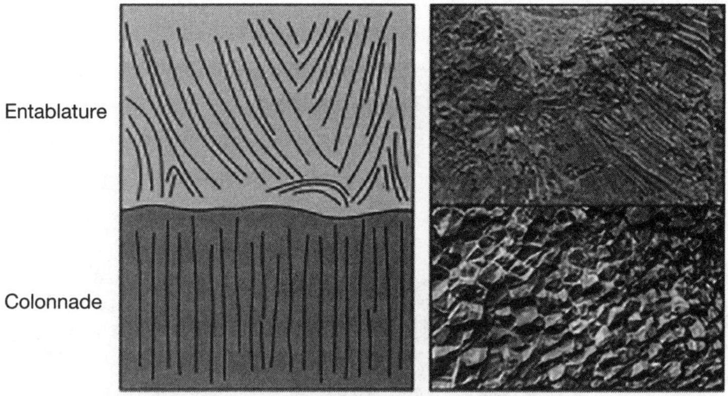

Figure 4.5 Entabular versus colonnade columnar basalt. Modified from Spry (1962) and Jovanelly (2020).

result of slow cooling from the base upward and rapid cooling from the top downward. Perhaps the most famous columnar basalts are those of the Giant's Causeway found in Northern Ireland (55.2408, –6.5114). Here, an intense volcanic episode occurred 60 MYA, leaving behind 40,000 columns that were up to 12 m tall.

The terms "subglacial" and "intraglacial" classify volcanic eruptions by the landforms created by such events. Specifically, subglacial events are only those processes occurring under ice cover without direct contact with the atmosphere. The term "interglacial" is more general, since it also applies to eruptions and volcanoes that have broken through glaciers. Subglacial eruptions from a central vent can lead to the formation of tabletop mountains (also known as tuyas) or möberg ridges (also known as tindars) (Figure 4.6). While challenging to discern, tuyas typically have a much broader lava cap that is missing from tindars. Both have a broad base made of hyaloclastic lava. Hyaloclastic lava is hydrated tuff-life breccia rich in volcanic glass and has the appearance of angular flat fragments sized between a few millimeters to a few centimeters. This fragmentation occurs by the force of the volcanic explosion or by

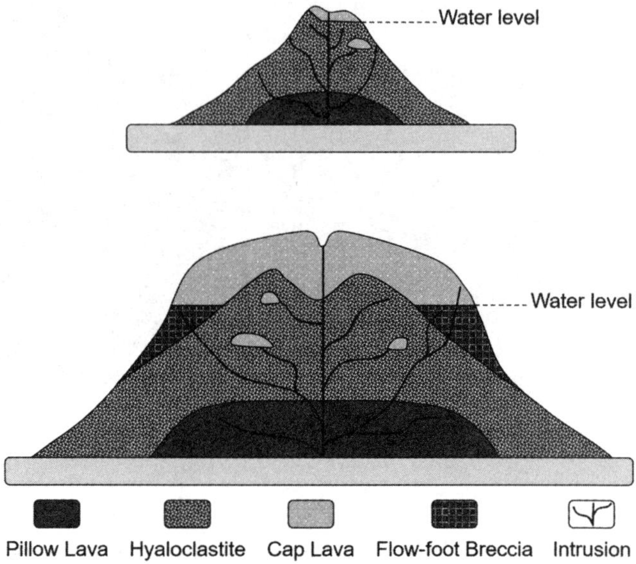

Figure 4.6 Comparison between tindar (top) and tuya (bottom) subglacial structures. Modified from Jakobsson and Johnson (2012) and Jovanelly (2020).

the thermal shock during rapid cooling. Subglacial-eruption pillow basalts are formed if the ice-to-water pressure is adequately high. Under low-pressure conditions, magma will immediately cool due to the rapid heat transfer from ice to magma, resulting in glass mineral textures like obsidian (Figure 2.6 E). While the creation of tuyas and tinder is rare worldwide, it is common in Iceland, as the island was completely covered in an ice sheet 2.6 MYA. Identification of these features not only confirms the presence of glacial ice but also reveals its minimum depth in situ. A textbook example near Vatnajökull National Park is a tuya called Herdubreid (64.1788, –16.3472).

VOLCANIC HAZARDS

5

WHAT ARE VOLCANIC HAZARDS THAT RESULT FROM VOLCANIC ACTIVITY?

How are volcanic hazards and risks defined?

There are approximately 1,350 active volcanoes on the planet, and about 500 of those have erupted in historical time; each year there are around 50–70 volcanic eruptions. Moreover, about 10% of the world population lives near a volcano. Thus, understanding volcanic hazards, along with the risk of living near one, has become increasingly important as population growth continues into the 21st century.

For this reason, volcanologists look at an active volcano in terms of what physical hazards it could produce and what would be the associated risk to people, property, or production should an event occur (Rymer, 2000). The resulting hazards and risks will depend on six factors:

1. Volcano type. The type of volcano is an important consideration, because shield volcanoes have less viscous magma that manifests in less explosivity than stratovolcanoes, for example (Figure 3.2). Not to say that shield volcanoes are not dangerous, but a gentle eruption at Kilauea, Hawaii (1,222 m; 19.5180, −155.0313), will likely create less overall destruction than a cataclysmic eruption at Mount Pinatubo, Philippines (932 m; 15.1436, 120.3500).

2. Recurrence interval. The recurrence interval, or repose period, of a volcano will determine how long a volcano has been dormant. The longer the period of quiescence, the more likely a larger eruption will occur because more magma and gases are being held within the magma chamber. Some volcanoes have a repose period of less than 10 years and generate nonviolent eruptions, while others can be dormant for hundreds to thousands of years, creating violent eruptions.
3. Geographical location. The geographic location, specifically which type of plate boundary a volcano is on, will dictate the volcano type and eruption style. Approximately 80% of active volcanoes on the Earth's surface are located within the convergent plate boundary that outlines the Pacific Ocean, referred to as the Pacific Rim of Fire (Figure 1.5). These volcanoes, usually classified as stratovolcanoes (Figure 3.2), are highly explosive due to their felsic magma sources (Figures 2.3 and 2.4). The other 20% of volcanoes are associated with plate divergence or hot spot activity, and they are considered to be much less of a global, or even local, hazard because mafic magma is of lower viscosity, which contains less volatile energy overall.
4. Effects of local climate. The local climate will determine variables like wind direction and speed. Explosive eruptions may eject a large volume (0.1 km to hundreds of kilometers) of ash into the atmosphere that may be carried great distances. For instance, a 17-hour Plinian-style eruption of Askja, Iceland (1,516 m; 65.0330, –16.7830), on March 29, 1875, dispersed ash 1 mm thick as far as 1,100 km away, in Scandinavia (Carey et al., 2010).
5. Time of year. Depending on the latitude, seasonality influencing temperature and precipitation patterns can have a big impact on an eruptive event. That is, an eruption occurring with the presence of water or ice will be intensified. Next, with ample water, secondary hazards can ensue, like lahars or jökulhlaups.

6. Population density. Basically, the more people living in the shadows of a volcano, the greater the risk of fatality. By way of explanation, the city of Yogyakarta, Indonesia, can be used. One of the most densely populated areas in the world, this city of nearly 5 million people is situated within a 30-km radius of the active stratovolcano called Merapi (2,910 m; –7.5401, 110.4460). That being so, a large eruption could have devastating consequences to those living on this island.

What is considered a volcanic hazard?

Hazards that stem from volcanic activity produce direct and indirect consequences. Those considered a direct hazard institute death or injury to humans or wildlife, or destroy property. Indirect hazards have implications to the natural environment that can later manifest in famine, habitat degradation, and climate change. These hazards include tephra and gas emissions, pyroclastic fall and density currents (i.e., pyroclastic flows), lahars, jökulhlaups, debris avalanches, and lava flows (Figure 5.1).

Type	Description	Risk
Tephra emission	Small particles of volcanic rock emitted into the atmosphere	• Respiration problems for some individuals • Significant climate cooling leading to crop failure and famine • Damage to aircraft
Gas emission	The emission of gases before, during and after an eruption	• Climate cooling leading to crop failure and famine • In some cases, widespread poisoning
Pyroclastic fall	Vertical fall of tephra in the area surrounding an eruption	• Thick tephra coverage of areas close to the eruption (1Km to 10's of Kilometers) • Collapsed roofs
Pyroclastic density current	A very hot (several 100°C) mixture of gases and volcanic fragments (tephra) that flows rapidly (up to 100s of kilometers per hour (km/h) down the side of a volcano	• Extreme haard – destroys anything in the way
Lahar	A flow of mud and debris down a channel leading away from a volcano, triggered either by an eruption or a severe rain event	• Severe risk of destruction for anything within the channel – lahar mud flows can move at 10's of Km/h
Jokulhlaup	A glacial outburst flooding event	• Catastrophic flooding and movement of sediment that is either deposited or eroded
Debris avalanche	The failure of-part of a volcano, either due to an eruption or for some other reason,leading to the failure of a large portion of the volcano	• Severe risk of destruction for anything in the path of the debris avalanche
Lava flow	The flow of lava away from a volcanic vent	• People and infrastructure at risk, but lava flows tend to be slow (less than Km/h) and are relatively easy to avoid

Figure 5.1 A list of important volcanic hazards with their associated event description and risk.

Once a volcanic eruption occurs, volcanic ash (tephra) and gases are ejected as a column into the atmosphere. The tephra can fractionate, or become smaller in size, as it is transported through the air, allowing for it to be carried great distances. However, when the tephra falls directly to the surface covering the volcanic region, it is referred to as a pyroclastic fall deposit. In terms of the gases released, approximately 99% of the emission by a volcano are water vapor (H_2O), carbon dioxide (CO_2), or sulfur dioxide (SO_2) (Figure 5.2). The remaining 1% can include other dangerous components like hydrogen sulfide (H_2S), carbon monoxide (CO^-), hydrogen chloride (HCl^-), and hydrogen fluoride (HF^-).

Either concurrently or subsequently, a volcanic eruption can generate abrupt movement of ash and gases downslope combining to form a density current, also referred to as a pyroclastic flow. By definition, a pyroclastic flow is a dense, fast-moving flow of solidified tephra pieces, ash, and hot gases. There are three common ways a pyroclastic flow can form. The first is when the ejected column of ash and gases loses its

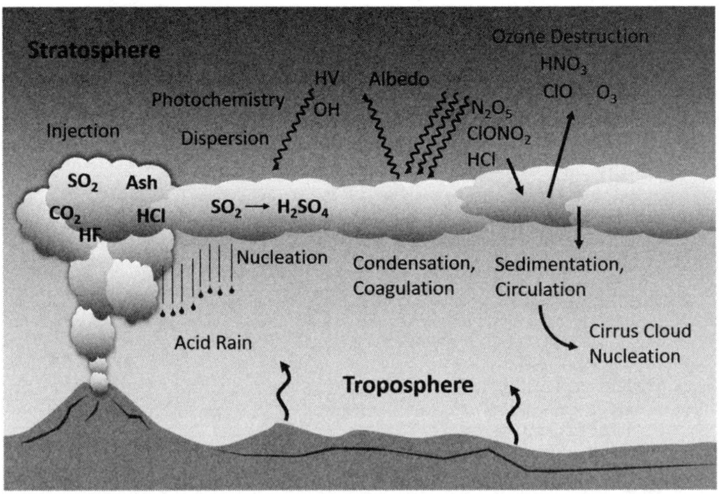

Figure 5.2 Cycling of volcanic gases in stratosphere. Modified from McCormick et al. (1995).

upward momentum and falls back to the ground surface. The second is when the volcanic material is ejected as a lateral blast (e.g., Mount St. Helens 1980 eruption). The third is when a lava dome over-steepens and becomes mobilized, often without warning (Mount Pelée, Martinique, 1902 eruption) (Figure 5.3). A pyroclastic flow can reach more than 800°C, burning anything it its path and it can move upward of 700 km/h.

Like pyroclastic flows, lahars can also occur during or after a volcanic event in areas of steep terrain. However, lahars only form in situations when large amounts of water become available, like during torrential rains brought on by a monsoon. The accumulation of water on the volcanic flank will create a flow deposit that incorporates additional sediment with continued movement. Lahars can also travel great distances (>100 km) and at great speeds (80 km/h). It is not uncommon for lahars to experience a "snowball" effect, where forceful downslope movement continues to accumulate material. Moving at rates between 10 and 10^7 m^3/s, lahars have the capacity to increase their volume to more than 10X their initial size before losing velocity with decreasing slope steepness. Following gravity, the thickest deposits are found in low-lying areas at the base

Partial List of Fatalities Associated with Dome Building, Collapse, and Destruction in the 20th Century

Volcano	Date	Fatalities
Mt. Pelée (Martinique)	1902	29,025
Soufrière (St. Vincent)	1902	~1,600
Santiaguito (Guatemala)	1929	~1,000
Merapi (Indonesia)	1954 1976 1994	64 28 66
Mt. Unzen (Japan)	1991	44
Soufrière Hills (Montserrat)	1997	19

Figure 5.3 Partial list of fatalities associated with dome building, collapse, and destruction in the 20th century. Modified from Fink and Anderson (2015).

of volcanic peaks, which often make for suitable homesteading locations.

A debris avalanche is similar to a lahar, but without the water. In some instances, a debris avalanche can turn into a lahar. Likely, this was the scenario that influenced the prehistoric Oscela Mudflow from Mount Rainer, Washington, that filled valleys with deposits over 200 m thick, with flows that spread over 120 km^2 and continued to travel 20 km under the water of Puget Sound. It inundated an area of 350 km^2 (Vallance, 2000).

Although rare, volcanism underneath a glacier can create a flooding event referred to as a jökulhlaup. In this situation, the caldera of the volcano will act as a basin for the meltwater that is produced by the heat from the upwelling magma. Once the caldera is full, the meltwater will breach and flow down the volcano's flanks, creating a sudden flooding event. "Jökulhlaup" is an Icelandic word, which speaks to their occurrence in this country.

Lava flows are the most common volcanic feature on Earth. They are the result of the outpouring of magma onto the Earth's surface. Despite being produced by effusive eruptions that are nonviolent in nature, lava flows can have high discharge capacities (10–1,000 m^2/s) and, spreading up to 50 km in length (Figure 5.4), can quickly overrun human settlements.

Composition	Typical length (km)	Mean thickness (m)	Mean discharge rate (m^3s^{-1})	Flow (& flow field) volume (km^3)
SiO2 < 55 wt% (basalts, basaltic andesites)				
	<10	3-20	10-100	0.01-0.1
	(basalts to 50)		(basalts to 1000)	(<1-2)
For example, Kilauea & Mauna Loa, Hawaii; Etna & Vesuvius, Italy; Iceland; Piton de la Fournaise, Réunion Is; Lanzarote & Tenerife, Canary Is.; Arenal, Costa Rica; Paricutin, México.				
SiO2 > 55 wt% (andesites, dacites, trachytes)				
	<5	20-300	1-10	0.01-1.0
	(some to I5)		(andesites to 100)	(<10-20)
For example, Lonquimay, Chile; Nea kameni, Santorini, Greece; Hibok-Hibok, Philippines; Trident, Alaska.				

Figure 5.4 Major historic lava flows. Modified from Kilburn (2000).

How has the collapse of lava domes resulted in the deadliest volcanic hazard for the 20th century?

First, lava domes form when viscous magma piles up near the vent of a volcano (Figure 1.4). Although inconspicuous, during the building of a lava dome, gas pressure can accumulate in the subsurface conduit. This pressure can destabilize the structure and build to an explosive eruption that can cause massive devastation by generating pyroclastic flows. The explosion or collapse of lava domes ranks among the deadliest natural disasters recorded in the 20th century (Figure 5.3). The most tragic event occurred on May 8, 1902, when a lava dome collapsed at Mount Pelée on the Caribbean Island called Martinique (1,208 m; 14.8102, –61.1633). This event generated a single pyroclastic flow current that was responsible for claiming the lives of more than 29,000 people and destroying the city of Saint Pierre. More recently, in 1991 the collapse of a lava dome on Mount Unzen, Japan (702 m; 32.7797, 130.2677), triggered a pyroclastic flow that traveled 4.5 km away from the crater, causing 44 fatalities. Following, in 1997 an explosive eruption triggered a massive dome avalanche on a Caribbean island called Montserrat (584 m; 16.7105, –130.1755). This forced half of the island's population to evacuate and resulted in 19 causalities.

Why are lahars so dangerous?

Lahars, or mudflows, are considered to be one of the destructive natural disasters because they quickly, and unexpectedly, inundate communities in distal areas that are great distances away from volcanic peaks (Figure 5.5). The deadliest lahar in history was recorded on November 13, 1985, when the glaciated Nevado del Ruiz, Colombia (5,279 m; 4.8920, –75.3241), erupted, sending lahars down four major drainage channels. Approximately 74 km east of the summit, the town of Armero was met with a 40-m-tall wave, killing 28,000 residents. In other active volcanic areas, like Mount Rainer, Washington

Lahar date	Specified locations	Lahar trigger	Distance of specified location from lahar source (km)	Travel time from lahar source to specified location (min)	Average speed (meters per second)
1926	Points along Hurano River, downstream of vent at Tokachidake volcano, Japan	Eruption (pyroclastic density current on snow and ice)	2.4 6.4 20.0 23.2	1 4 17 29	42 26 16 5
1980	Points along Pine Creek, downstream of vent at Mount St. Helens, USA	Eruption (pyroclastic density current on snow and ice)	10.1 14.1 16.8 20.9	7 11 13 19	24 16 18 14
1985	Point along Denjo River, downstream of landslide at Mount Ontake volcano, Japan	Earthquake-triggered slope failure	11.4	10	19
1985	Points along Río Chinchiná, downstream of vent at Nevado del Ruiz volcano, Colombia	Eruption (pyroclastic density current on snow and ice)	11.3 33.0 68.6	20 70 172	10 7 6
1990	Points along Drift River, downstream of vent at Redoubt Volcano, Alaska	Eruption (pyroclastic density current on snow and ice)	7.6 18.5	8 27	16 10
1994	Points along Río Páez, downstream of vent at Nevado del Huila volcano, Colombia	Earthquake-triggered slope failures	4 9 30	2 to 3 6 to 9 20 to 30	22 – 33 17 – 25 17 - 25

Figure 5.5 Examples of lahar travel times from lahar source areas (points of initiation) to selected locations in downstream river valleys. Modified from Pierson et al. (2014). Travel times are partly a function of lahar magnitude and are determined from eyewitness accounts, instrument recordings, and back-calculations of flow velocity based on physical evidence.

(4,392 m; 45.8533, −121.7605), geologists have mapped 60 different lahar deposits that have been generated over the past 10,000 years. To date, there are ~300,000 people who live and work in buildings located on top of these deposits; this indicates the urgency for engineered protection structures, lahar warning systems to enable evacuation, and effective response to and recovery from lahars when they do occur (Pierson et al., 2014).

Is the collapse of lava domes considered to be a hazard?

Although inconspicuous, during the building of a lava dome, gas pressure can accumulate in the subsurface conduit. This can destabilize the structure and build to an explosive eruption that can cause massive devastation, including pyroclastic flows. Deaths associated with the explosion or collapse of lava domes are among the deadliest natural hazards recorded for the 20th century (Figure 5.3). The most tragic event occurred on May 8, 1902, when a lava dome collapsed at Mount Pelée on the Caribbean island called Martinique (1,208 m; 14.8102, –61.1633). This event generated a single pyroclastic flow current that claimed the lives of more than 29,000 people and destroyed the city of Saint Pierre. More recently, in 1991, the collapse of a lava dome on Mount Unzen, Japan (702 m; 32.7797, 130.2677), triggered a pyroclastic flow that traveled 4.5 km away from the crater, causing 44 fatalities. Following, in 1997, an explosive eruption triggered a massive dome avalanche on a Caribbean island called Montserrat (584 m; 16.7105, −130.1755). This forced half of the island's population to evacuate and resulted in 19 causalities.

Are the gases emitted during a volcanic eruption life-threatening?

When there is a volcanic eruption, it is not uncommon to see a billowing plume of smoke directly above the vent. This process of volcanic outgassing releases ash and volatile gases that were once dissolved in the magma into the atmosphere. The most abundant gases emitted during a volcanic eruption include water vapor (H_2O), sulfur dioxide (SO_2), and carbon dioxide (CO_2), with lesser amounts of carbon monoxide (CO^-) hydrogen sulfide (H_2S), hydrogen fluoride (HF^-), hydrogen chloride (HCl^-), and hydrogen bromide (HBr^-) (Figure 5.2).

Of these gases, water vapor, seen as steam, will be predominant, representing at least 80% of the combined release, and poses little harm other than that it can carry tiny particulate matter called aerosols. However, the other two major gases found in volcanic eruptions, SO_2 and CO_2, can cause direct harm. When SO_2 is released from a volcano, it begins to interact with water and sunlight to become sulfuric acid (H_2SO_4). This produces a visible smog (vog) that can cause respiratory problems if inhaled, and can irritate the skin, throat, and eyes. Eventually, the atmosphere will release the vog as damaging acid rain that will cause indirect challenges to human health by diminishing the harvest of crops, for example. Some volcanoes, like Kilauea, Hawaii (1,222 m; 19.5180, −155.0313) are known to produce vast amounts of SO_2 gas during sustained eruptions (between 500 and 14,000 metric tons) and are monitored closely for this reason.

Likewise, the inhalation of CO_2 (and CO^-) gases can be dire. Both, CO_2 and CO^- are slightly denser than air, permitting them to collect in low-lying areas. These gases are referred to as the "silent killers" because they go unnoticed as they are odorless, colorless, and tasteless. Symptoms of mild exposure can include headache and drowsiness, but long-term or high exposure can lead to death. Among the worst volcanic-related CO_2 and CO^- exposures occurred in 1986 at a crater lake called Lake Nyos, Cameroon (2,762 m; 6.4360, 10.2980), when, without warning, the lake exploded, killing almost 1,800 people and 3,500 livestock. It is estimated that 300,000 tons of CO_2 was released from the stagnated lakebed, creating a gas cloud that spread over a 25-km radius.

The presence of hydrogen sulfide (H_2S) is noticeable in volcanic regions because it smells like rotten eggs. Often, H_2S is linked with geothermal areas that contain hot springs, geysers, or bubbling mudpots, because the chemical is formed when sulfur gases from the magma mix with groundwater. Moreover, the release of H_2S can be used as an indicator of volcanic dormancy because the pace at which the sulfur gases are being released is slow enough that they can be absorbed by the water column instead of being immediately released at the surface as SO_2.

At low levels, this sulfur derivative can cause eye irritation, and at high concentrations (>0.01%) H_2S can cause upper respiratory problems and pulmonary congestion. The inhalation of more than 500 ppm H_2S can cause unconsciousness and death. However, these levels are typically only found nearest the vent and just before eruption activity.

A group of gases called hydrogen halides include hydrogen fluoride (HF^-), hydrogen chloride (HCl^-), and hydrogen bromide (HBr^{-}) and can be released in some quantities during a volcanic eruption. These gases are highly soluble and can quickly incorporate themselves into volcanic plumes or attach to particulate matter such as ash. Once in the atmosphere, hydrogen halides can fall to the surface as acid rain or acidic outfall. This can directly result in the poisoning of groundwater, surface water, and soil, which later results in famine.

Are jökulhlaups predictable?

A glacial outburst flood (jökulhlaup) can occur anywhere in the world where there is a glacier capping an active volcano, but they are most common in Iceland, where there is both ice and a magma hotspot. A jökulhlaup materializes when the heat emitted from magma moves up a central vent (Figure 1.4) to warm the base of the glacier, creating meltwater. The meltwater accumulates as a subglacial lake within the volcano's crater, which is now acting as a catchment basin. Once the crater is at its capacity with meltwater, the flanks will breach and a flooding event will occur.

These types of flooding events can be particularly dangerous because they can take place prior to, during, or in the absence of a volcanic eruption. However, there are a few precursory warning signs that are monitored. First, in Iceland, geologists have a good understanding of the size, shape, and volume of the craters hiding underneath the ice. If this is known, then a storage capacity can be calculated. Next, the rate at which meltwater forms within the crater is measured. If geologists know the volume of the crater, and the rate of infilling, then they can make some educated guesses as to when a breach may occur. A final indicator that a flooding event may happen soon includes the formation of a depression on the surface of the icesheet that can be viewed aerially.

A large jökulhlaup occurred in Iceland in 1996 at a volcano called Grímsvotn (1,342 m; 64.4169, −17.3219). Using the

Figure 5.6 Aerial photograph of the world's largest sandur plain, Skeidarársandur, Iceland, that was created by jökulhlaup events over the last 10,000 years. Photo credit: Alamay.

forementioned precursors, the Icelandic Meteorological Office was able to evacuate the area prior to the flooding and subsequent eruption. The upper limit of flow was reported at 45,000 m^3/s, which carried icebergs the size of houses that slammed into the Skeidará Bridge (63.9558, −17.4647), causing it to collapse. While no lives were lost, the coastline was dramatically reshaped through erosional and depositional processes, as it had been in the past. In Iceland, flooding events over the past 10,000 years have produced enough sediment to create the world's largest sandur plain (Skeidarársandur, 1,000 km^2; 63.9487, −17.4133), which extends from the sea 25 km inland (Figure 5.6).

The understanding of the past recurrence of jökulhlaup events in Iceland is derived from the interpretation of sediment that is deposited once the water dissipates in velocity. By measuring the thickness and extent of these deposits, geologists can approximate the intensity of each event. For example, Tomasson (1996) estimated the velocity of the flooding discharge during a 1755 event at Katla, Iceland (1,377 m; 63.6469, −19.1302), to be ~300,000 m^2/s or equivalent to the flow volume of the Amazon River. He also estimated that the coastline expanded ~4 km toward the ocean, although it has since eroded.

How is it possible to generate a tsunami wave by volcanic activity?

First, what is a tsunami? A tsunami is a catastrophic wave that is generated by submarine earthquakes of $M \geq 7.0$, underwater landslides, volcanic eruptions, or on rare occasions, meteorite impact. From the point of initial disruption, the tsunami waves will radiate out from the epicenter at equal distance and speed until stopped by land or infrastructure and will lose velocity with distance and time.

The intensity of a tsunami is characterized by two variables (Figure 5.7):

1. Wave height, also called run-up. This is the measurement of the maximum vertical height of a tsunami wave when it reaches land. The height is dependent on the proximity to the epicenter or disruption origin, the bathymetry, and the local topography. The run-up height is usually not taller than 30 m above sea level, although there are exceptions.
2. Wave speed. The travel time of a tsunami wave correlates to the depth of water it is moving through. If a wave is moving through a deep ocean, it will not interact with the seafloor. Thus, the wave can move up to 800 km/h because friction is not impeding its speed. Additionally, in the deep ocean the wavelength can be up to 1.6 km long, whereas the wave height may be less than 10 cm. The forward momentum carried by the wavelength will be transferred to wave height when approaching shallow waters, which will lessen its speed to 30–50 km/h.

Location	Age	Cause of tsunami	Wave height	Travel distance	Impact of tsunami
Santorini Mediterranean Sea	1638 B.C. (?)	Explosive eruption, caldera formation Pyroclastic flows into sea	10-15 m	150 - 500 Km ?	Destruction of Minoan civilization on Crete
Unzen, Japan Ariake Sea	1782 A.D.	Debris avalanche into sea	10-55 m	20 - 50 Km	More than 10,000 fatalities
Tamboro, Indonesia Java Sea	1815 A.D.	Pyroclastic flows into sea	>10m	> 100 Km	More than 10,000 fatalities
Karakatoa, Indonesia Sundra Straits	1883 A.D.	Explosive eruption, caldera formation Pyroclastic flows into sea	5-35 m	800 Km	More than 36,000 fatalities
Augustine, Alaska Cook Inlet	1883 A.D.	Debris avalanche into sea	7-9 m	> 100 Km	Damage to coastal villages, no known fatalities
Mt. St. Helens, Washington Spirit Like	1980 A.D.	Debris avalanche into lake	260 m	4 Km	No known fatalities
Lake Nyos, Cameroon	1986 A.D.	Exhalative emission of CO_2	25-75 m	5 Km	No known fatalities from tsunami
Karymskoye, Kamchatka Karymskoye Lake	1996 A.D.	Phreatomagmatic eruptions	30 m	5 Km	No known fatalities

Figure 5.7 Comparison of impactful volcanic tsunamis. Modified from Begét (2000).

Tsunamis sourced by volcanism are not uncommon. Over the past 250 years there have been 90 tsunamis with volcanic origin (Begét, 2000). This is probably because of the ~1,350 active volcanoes in the world, approximately 90% are located along the subduction zone known as the Pacific Ring of Fire (Figure 1.5). Moreover, this active subduction zone that hosts the majority of the Earth's volcanoes outlines a significant portion of the Pacific Ocean, covering ~40,000 km^2 (Figure 1.6). And there are an estimated 50,000 seamounts (underwater volcanoes) that cover the seafloor (Figure 1.7). However, the large possibility of tsunami wave generation in the ocean should not reduce the chance of such an event occurring within a lake system where appropriate conditions occur.

Volcanoes can initiate tsunami waves either directly from explosive activity or indirectly through secondary consequences or aftermath. Direct mechanisms include volcanic eruption, eruptions of undersea volcanoes (seamounts), phreatomagmatic explosions, and caldera collapse. Indirect mechanisms include pyroclastic flows, lahars, debris avalanches, or landslides entering the sea (Figure 5.1). That is, tsunamis can be generated concurrent to the eruptive activity or after, but regardless, both situations are known to result in death (Figure 5.1).

With either mechanism, the transfer of energy from on-land to water exacerbates the threat of the volcanic episode and extends the consequences well beyond the flanks of the volcano. About 5% of all historic tsunamis, including some of the most destructive tsunami waves ever seen on Earth, were produced at volcanoes by eruption or associated processes (Begét, 2000). Two devastating historic volcanic tsunami events resulted from the eruptions of Mount Unzen, Japan, 1792 (698 m; 32.7811, 130.2672), and Krakatoa, Indonesia, 1883 (463 m; –6.1005, 105.4250) (Figure 5.7). On May 21, 1792, Japan experienced its worst natural disaster related to volcanic eruptions at Mount Unzen. After four months of activity, the collapse of a lava dome spurred a debris avalanche down its eastern flank, meeting the Ariake Sea. This generated a megatsunami—a wave 55-m-tall that resulted in 14,000 deaths. Next, on August 27, 1883, the collapse of Krakatoa triggered a pyroclastic flow that sent a 35-m-tall tsunami wave into and away from the Sundra Straits that would overrun small islands and lead to a death toll of 36,000.

Are lava flows considered to be a hazard?

Lava flows are the most common volcanic features found on Earth. However, they rarely result in human death because of the low viscosity of the slow-moving mafic magma that solidifies as basalt (Figure 2.6 C). In fact, lava flows are often coined "tourist eruptions" because they can be experienced in close proximity with little danger if eyewitnesses pay attention to posted signs and warnings issued. Although much less likely to be dangerous than eruptions at stratovolcanoes, the close observation of lava flow fields in shield volcano settings has resulted in spectator death (Figure 3.2). The main causes of death are attributed to unforeseen phreatic eruptions, lahars, pyroclastic flows, or the presence of elevated toxic gases.

Additionally, lava flows can be considered hazardous because in most scenarios, the rock cools first on the exterior. The color of the lava at the surface is a predictor of its temperature; the darker the color, the cooler the lava (Figure 3.1). However, although the lava may appear solidified, the inside may still be molten. This is why walking on new lava flows is discouraged.

While danger to human life is low, lava flow fields have been seen to cause great harm to communities living near them. An example is the 1973 eruption of Eldfell (107 m; 63.4322, –20.2494), a volcano on Heimaey Island in the Vestmannaeyjar archipelago near Iceland. When this eruption occurred, the lava flowed down the flank of the volcano toward the city center.

The military pumped water from the ocean to cool and redirect the lava. Although this was quite successful, 400 buildings in the nearby fishing port town were destroyed.

How are aircraft affected by volcanic eruptions?

Jet airliners, flying at more than 800 km/h with hundreds of passengers, can be impacted by both volcanic ash and gases emitted from eruptions as far away as 3,000 km. The challenge faced by the airline industry is that most of the volcanoes that have produced hazardous ash clouds in the past several decades are located on the Pacific Rim, which dominates flight patterns and hosts some of the world's busiest international airports (e.g., Tokyo, Hong Kong, Shanghai). Because volcanic ash is made up of tiny shards of hard, angular silica grains, it can immediately wreak havoc on the external and internal components of a plane, thereby reducing the pilot's visibility, abrading exterior components, and causing engine failure. Moreover, volcanic ash clouds are not detectable by aircraft radar, so complete avoidance is often impossible.

After many near fatal airline accidents and interruptions to air traffic resulting from volcanic activity that occurred between 1980 and 1991 (i.e., eruptions of Mount St. Helens; Galunggung, Indonesia; Redoubt, Alaska; Mount Pinatubo), the first International Symposium on Volcanic Ash and Aviation Safety convened. This motivated the interdisciplinary cooperation of global airports, volcanologists, air traffic controllers, and governments to develop warning systems for volcanic activity.

While the warning system has proven successful, there are many other consequences of volcanic eruptive events relating to air travel and economics. A more recent case study is the eruption of Eyjafjallajökull, Iceland (1,515 m; 63.6344, −19.6102), from April 14 to 21, 2010. This 7-day eruption produced an ash column 13 km tall that was carried east, across the North Atlantic, toward Europe. This event grounded 100,000 flights, stranded 10 million passengers, and cost the airlines approximately $1.7 billion in revenue. While inconvenient, this volcanic eruption is considered small in comparison to those found in the geological record of many Icelandic volcanoes and others along the Pacific Rim.

EROSION

6

HOW DO VOLCANIC LANDSCAPES CHANGE WITH TIME?

Are volcanoes permanent landscape features?

Like all landscape topography, volcanoes are subject to weathering and erosional processes that destroy their structure over time. Weathering can be defined as the physical or chemical breakdown of the rock. The most common forms of physical weathering include the following:

1. Gravity, or simply the movement of rocks downslope.
2. Frost wedging, which occurs when fractures within the rock become in-filled with water that freezes, expands, and further cleaves the rock.
3. Exfoliation, which involves the processes of releasing pressure off the rock when weight is removed causing expansion and uplift, such as the retreat of a glacier exposing the underlying bedrock.
4. Salt weathering, an occurrence found in marine environments where salt from evaporated seawater precipitates in cracks to cause further decay.

Often, chemical weathering acts in conjunction with the physical weathering processes to exacerbate the degradation of volcanic landscapes. Chemical weathering is the breakdown of the mineralogical components of the rock by water, oxygen, or carbon dioxide. In terms of volcanic landscapes, there are three main types of chemical weathering:

1. Oxidation, or the process that results when oxygen reacts with metal elements in rocks to form oxides, producing rust, for example.
2. Clay formation, or the disintegration of water-soluble minerals within a volcanic rock, like plagioclase, producing a clay deposit.
3. Hydrothermal alteration, the upwelling of hot (100°C–400°C) hydrothermal fluids containing volatile components like water, carbon, sulfur, and chloride salts that can decompose rock into clay, replace minerals, or deposit ore minerals. Hydrothermal alteration can result in changes to the rock's color, porosity, permeability, and texture.

While volcanic landscapes are being dismantled into smaller and smaller units via weathering, processes of erosion are working to transport the clasts away from the source by the movement of wind, water, or glacial ice over a surface. Sometimes, a volcano can initiate drastic mechanisms of erosion through volcanic tremors, conduit wall collapse, or pyroclastic flows, thereby moving large amounts of material at once. Other times, rates of erosion can vary wildly, depending on the texture of the rock. That is, if the rock is in the form of hardened basalt (Figure 2.6 C), it will take much longer to erode than unconsolidated tephra (Figure 2.6 L). Likewise, the influence of climate, particularly tropical climates hosting hot temperatures and abundant rainfall, will expedite both processes. In fact, erosion processes on active volcanoes in humid climates result in some of the highest sediment yields on Earth (Thouret et al., 2014).

Ultimately, the longevity of a volcano against weathering and erosion will be determined by the volcano's morphology, that is, its size and shape (shield, strato, cinder, etc.) (Figure 3.2). The oldest known volcanic deposit on Earth is located along the Hudson Bay in a small village called Inukjuak, Canada (32 m; 58.4588, –78.1061). The lava flows here have been dated to be ~3.825 billion years old.

How quickly does volcanic rock weather and erode? And what are the byproducts?

Basalt covers approximately 60% of the Earth's surface, making it the most predominant rock type found on our planet. The overall rate of weathering and erosion of basalt is estimated to be between 0.1 and 0.5 cm/yr. Although considered a slow process (geologically speaking), decay can increase with exposure to water, wind, or glacial ice. Additionally, these processes will vary depending on whether the rock was initially crystalline, porous with vesicles, or unconsolidated (e.g., tephra).

Where there is erosion, there can also be deposition. And it is not uncommon for volcanic islands to form depositional features, like beaches. These beaches are made of the gradual weathering and erosion of the island's rock that is transported to the shoreline, where it is usually protected by a bay. However, in some instances, volcanic beaches can develop very quickly. A modern example of a rapid beach face expansion is the black sand beach at Punalu'u, Hawaii (19.1361, –155.5047). While Punalu'u Beach has existed for more than 2,000 years, contributions of new sand occurred from the eruptions of Mauna Loa and Kilauea over the past decade. The beach was expanded several times as hot effusive lava flowed down the volcanic flanks, directly meeting the ocean, where it instantaneously exploded into small pieces that expedited the formation of sand size particles.

Chemical weathering of basalt is important to the understanding of global climate, as it can promote the permanent storage of carbon dioxide (CO_2). A portion of the mineral composition of basalt is calcium and magnesium. When basalt dissolves in water, it releases these soluble minerals that react with CO_2 stored in the water column. This encourages a chemical reaction to occur, called carbon mineralization, that binds CO_2 in the process of forming stable carbonate minerals, like calcite and dolomite. While this is a natural process of storing CO_2 at the bottom of the ocean, it proves insignificant compared to the amount of CO_2 produced from anthropogenetic sources. However, scientists have been able to mimic the carbon mineralization process in Iceland and have discovered how to sequester CO_2 from the atmosphere and store it underground in vesicular basalt as calcite. This is referred to as the CarbFix Method (Matter et al., 2011).

How quickly can a volcanic island erode?

The best documented case of an island's growth and attrition is an island off of the southern coast of Iceland called Surtsey (155 m; 63.4277, –20.2674). On November 14, 1963, fisherman noticed smoke on the horizon. Soon after that, several Icelandic geologists flew over the North Atlantic Ocean to witness the birth of a new island. The growth of this island lasted for 3 years and was chronicled in detail by 85 site visits made by Sigur Thórarinsson (1964, 1966, 1968) and by innumerable scientists who contributed to the nine volumes of the Surtsey Research Progress Reports. Among many revelations, the Surtsey eruption provided insight into the relationship between water-magma mixtures and explosivity of the resulting event. The final building phase of Surtsey in June 1967 left the island at 1.3 km^2. This pristine landscape has been protected since its origin as a United Nations Educational, Scientific, and Cultural Organization (UNESCO) World Heritage Site and continues to be a site of scientific discovery. Moreover, this well-documented phreatomagmatic eruption has become formally classified by volcanologists as the "Surtseyan-style" eruption and has been applied to a number of other locations where this kind of activity has occurred (e.g., Taal, Philippines; Kavachi Volcano, Solomon Islands; Ambae, Vanuatu).

Monitoring the subsidence, consolidation, and erosion of Surtsey has been an ongoing process for more than 50 years. The initial ground surveys completed on Surtsey included the installation of 42 bench markers along a 1,600-m east–west transect, which was measured frequently with markers being added to or improved. The initial rates of subsidence measured in 1967–1968 were 15–20 cm per year and likely represent the compaction of volcanic material and the seafloor underneath the island. Overall, the rate of subsidence measured was 1.1 ± 0.3 m during the first 24 years (Moore et al., 1992) and 1–2 cm per year thereafter. Compaction is important to the discussion of Surtsey because the initial total porosity was reported as high as 40%–50%. As an example, it was estimated that by 2007, 95% of the volume of tephra was altered to compacted tuff (Jakobsson et al., 2009). Thus, in addition to the height reduction of Surtsey, erosive forces of temperature and coastal geomorphological processes have caused both subaerial and submarine destruction to reduce the area of the island to half its original size in 1967. Romagnoli and Jakobsson (2015) make comparisons of erosional and depositional changes that have altered Surtsey's shape over the past 50 years, while noting that extreme rates occurred during the first few decades and then slowed in recent times. Perhaps the most drastic reduction from the high-energy coastal processes has been that of the unconsolidated tephra coastline, which eroded hundreds of meters to produce cliff faces and a submarine shelf on the southern portion of the island. Here, a boulder beach has formed in the swash zone, which, in turn, protects both the remaining cliff faces and a submarine shelf on the southern portion of the island.

Based on all of the data collected at Surtsey to date, and assuming that the current rate of erosion does not change, it is estimated that the island will erode below sea level by 2100.

What does the inside of a volcano look like?

In the most basic terms, a volcano has a central vent that is connected to a magma chamber (Figure 1.4). During an eruption, magma from the chamber upwells to fill this vertical conduit before it is released at the surface as lava, tephra, ash, or some combination of all three. Toward the end of the eruption cycle there can be some magma of higher viscosity that remains in the vent, where it will slowly solidify. This intrusive solidification process permits time for the magma to release gases, which ultimately reduces the magma's porosity to create a dense crystalline lava inside the vent of the volcano. This crudely cylinder-shaped feature represents the volcano's internal plumbing and is called a volcanic plug. The rock type that persists will depend on the composition of the magma, but some of the most recognized volcanic plugs are products of basalt that solidify as columns (e.g., Devils Tower, Wyoming (1,174 m; 44.588, −104.6983). Additionally, it is not uncommon for the flanks of the volcanic plug to include xenoliths from the vent walls.

What happens after the plug solidifies within the vent depends on whether the volcano remains active or not. If the volcano is still active, then the plug can act as a barrier to the release of gases, causing the build-up of intense pressure, which can lead to an explosive Plinian-style eruption. During such an event, the plug itself will be destroyed. Otherwise, if a vent becomes blocked by a plug, the magma may push the plug to the surface as was observed at Mount St. Helens, Washington (2,479 m; 46.1910, −122.1952), in 2004. Or the magma may instead find another point of exit as was observed at Arenal, Costa Rica (1,590 m; 10.4633, −84.7030), during its 42-year eruptive phase from 1968 to 2010. Alternatively, if the volcano is not active, then its less durable exterior will begin to erode. Over time, the more resistant volcanic plug will become exposed, rising tall above its surrounding plain, as seen at Cabezon Peak, New Mexico (2,373 m; 35.6002, −107.0952).

What happens if a plug does not infill the vent during or after a volcanic episode? Usually, when a vent is no longer supported internally, it will become structurally unstable and collapse. However, in Iceland, a vent of an extinct volcano has been preserved and explored. Thrihnukagigur (515 m; 63.991, –21.6988) formed 3,500 years ago during a Strombolian-style eruption that created a crater at the surface through a central vent. Post-eruption, the magma chamber drained completely, leaving the vertical 200-m conduit empty. This unique and accessible vent has been of special interest to geologists who study geochemistry.

What happens when a resistant lava flow infills a soft-rock sedimentary basin?

Arid landscapes are often composed of easily erodible sedimentary rocks like sandstone and limestone; the American Southwest is a classic example of this. When volcanism occurs in desert environments, a unique geological phenomenon can occur called inverted topography. In the simplest terms, this type of relief inversion is a process where a negative landform becomes a positive one, for example, a valley becomes a ridge (van Wyk de Vries et al., 2022).

Inverted topography is commonly created when there is an effusive volcanic event within a basaltic monogenetic volcanic field. This occurs for two reasons. First, new basalt has a high porosity and permeability, allowing for ample drainage along the flow's edges. And second, lavas generally form positive relief. When lava begins to infill a preexisting valley that was carved by wind or water (Figure 6.1 A), the low slope of the valley makes it susceptible to continued deposition of lava over multiple volcanic episodes. The much harder lava caps and protects the underlying sedimentary rock from weathering and erosion (Figure 6.1 B). Over time, the adjacent area that is not capped will erode, forming high-standing mesas or

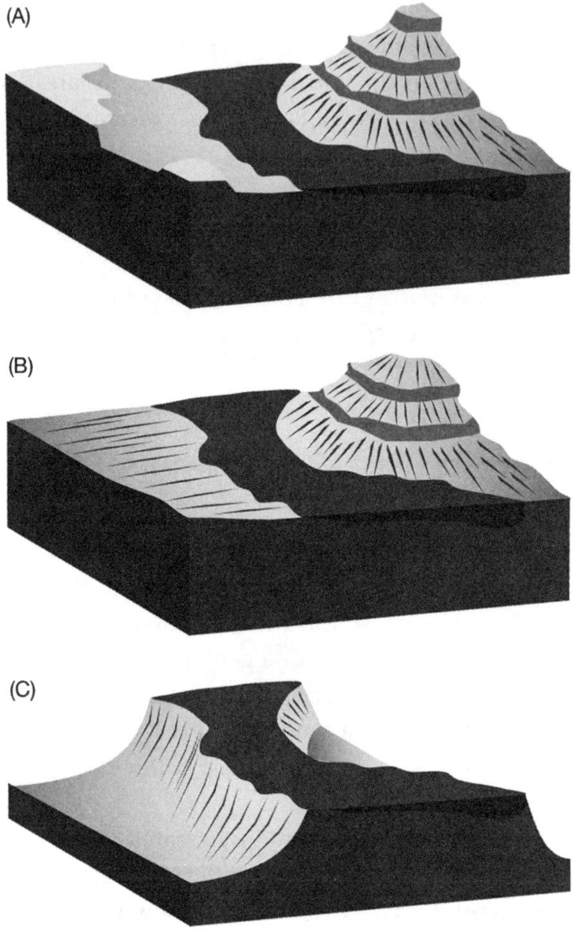

Figure 6.1 Illustration of the formation of inverted topography. A. Lava flows begin to fill in valley and low-lying topography. B. The hard volcanic rock caps underlying rock, preventing it from weathering and erosion. C. Erosion occurs around the resistant lava flows creating high-standing mesas above newly formed valleys. Modified from Muehlberger et al. (2005).

buttes above the newly formed valleys (Figure 6.1 C). This topography is found across the globe, including the western desert of Egypt, the Arabian Peninsula, the Atacama Desert, and in the Kutmag Desert in China.

Inverted topography is often used to interpret depositional environments using principals of stratigraphy. Typically, when an outcrop is approached, a geologist will examine the section from bottom to top, whereby the oldest layer is lower in the sequence and younger layers are consecutively higher. This method is a way to assign relative ages of rocks and is referred to as super position. With inverted topography, the oldest lava deposit will often be the taller mesas in a landscape, as time has allowed for deeper valleys to be cut around it. The younger flows will be progressively closer to the surrounding elevation.

Another volcanic setting that can produce inverted topography can be the formation of a lava lake within a preexisting tuff ring or maar. Here, a secondary eruption can infill the vent that acts as a basin for the effusive lava, which in this scenario, frequently solidifies as columnar basalt. The basalt is considerably more resistant to weathering and erosion than the unconsolidated pyroclastic material that the cone was made of. The remnant landform will be inverted hills like those of the Twin Peaks (Hopi Buttes, Arizona; 36.6913, –108. 8344) and Szent György Hills (Bakony-Balaton Highland Volcanic Field, Hungary; 46.8419, 17.4480).

In the 18th century, the ideology behind inverted topography was very important to understanding and supporting the foundation of geology as a science. To early geologists, the mechanism of creating inverted topography meant that erosion could occur over long periods of time, indicating that the Earth was significantly older than was previously thought (van Wyk de Vries et al., 2022). This concept became popular with the publication of the first geology textbook, *Principles of Geology*, by Charles Lyell (1833). Lyell himself studied inverted topography in Auvergne, France (45.7702, 3.3273).

How are dikes exposed at the surface?

In volcanic regions, dikes form when magma rises into a pre-existing crack or has enough energy to fracture the preexisting rock, thereby creating its own conduit before it solidifies in place. Dikes are usually vertical, or at some angle, and are discordant relative to the surrounding rock, allowing them to be used for relative dating. They can be singular, or present in multiples, then referred to as a dike swarm (Figure 6.2). For example, along a zone of structural weakness on a volcanic flank there may be hundreds of dikes like the occurrence at Sakurajim Volcano, Japan (31.5838, 130.6497; 992 m), in 2015 (Koike and Nakamichi, 2021). While dikes can vary wildly in size, from a few centimeters to 10 m wide, they are visible in a landscape because they will have a different color and texture than the surrounding rock.

Dikes are exposed at the surface through a process called differential weathering. Differential weathering occurs when you have multiple rock types in a location that decay at

Figure 6.2 Image of the "painted wall" in Black Canyon of Gunnison National Park, featuring linear white dikes in gray metamorphic rock. Photo credit: Alamay.

different rates. Usually, the dike will be more resistant to erosion than the surrounding rock, due to its more youthful age or composition. This prominence can be seen in an outcrop, or it can leave a visible ridge standing over low-lying topography. In New Mexico, there is a volcanic neck made of tuff called Shiprock (1,496 m; 36.7844, –108.6880). Seven dikes form a radial pattern around Shiprock that has an outcrop length of ~2,900 m, an average thickness of 2.3 m, and a maximum thickness of 7.2 m. These dikes are made of an igneous rock called minette, which is largely composed of the mineral's biotite and orthoclase. Minette is more resistant to weathering and erosion than the sedimentary sequences of siltstone and shale that use to cover the flat plain.

Can a hoodoo form in volcanic rock?

A hoodoo is a geologic formation that resembles a spire or chimney. Hoodoos are the result of weathering and erosion over extended periods of time. There are two commonalities identified across hoodoos:

1. They have interbedded layers that alternate between harder and softer rocks.
2. They have a hard cap layer that protects subsequent soft layers from erosion, thereby creating a pillar-like structure.

Although they are typically found in sedimentary rock, differential weathering of harder and softer igneous rock, like that of basalt and tuff, can result in a hoodoo. Hoodoos can range in height from 2 m to 50 m.

An example of these unusual landforms can be seen in Cappadocia, Turkey (38.6658, 34.8516). From 10 to 12 MYA a volcanic episode spread pyroclastic ash over an area of 20,000 km^2. Under high temperature conditions (>535°C), some layers of the pyroclastic material became welded together to form a rock called an ignimbrite. Initially, the area was a nearly horizontal plateau. Over time, the plateau became dissected along cooling fractures to form "fairy chimneys," as they are locally known. Different layers of ignimbrites have different resistance to erosion: softer layers (necks) are easier to erode compared to the harder ignimbrite caps. When the chimneys become isolated, the caps play an important role in slowing the further erosion of softer layers. At Cappadocia, the rate of erosion of the chimneys with caps is around 3.39 ± 0.36 cm/ky. Once the chimneys disappear, however, the rate of erosion increases significantly (28.0 ± 9.9 cm/ky) (Sarikaya et al., 2015). Similarly, the welded ignimbrites from the Turkey Creek caldera eruption 27 MYA deposited the material that would later be molded into the pinnacles at Chiricahua National Monument, Arizona (32.0058, −109.3563).

How does tafoni form?

Tafoni is a geologic structure that is often described as having the appearance of "swiss cheese" (Figure 6.3). Also known as honeycomb weathering, this structure can be found in coastal basalts that are exposed to seawater. The evaporated seawater will naturally penetrate through cracks, where it will precipitate the salt it carries. These tiny salt crystals exert pressure on the surrounding rock, thereby creating space and eventually carving small, circular cavities in the rock. The cavities are ellipsoidal or spherical in shape and can be < 1 cm to 10 cm across. This process can be enhanced if the basalt contains jointing or dikes, as there will be more areas for the saltwater to interface with the rock. Additionally, holes already contained within vesicle basalt will continue to grow in size with the saltwater intrusion. Varying size of the pits is the result of differential weathering and inconsistencies within the rock or decay process. A textbook example can be found in coastal India, at Korlai in the state of Maharashtra (18.5238, 72.9194). Tafoni in pyroclastic rock (tuff) has also been described on the

Figure 6.3 Photograph of tafoni formations in basalt rock. Photo credit: Thomas Griffith.

Isotake coast in Shimane, Japan (35.2286, 132.6416) (Kogure et al., 2022).

Are jökulhlaups erosive or depositional?

Both. A jökulhlaup is a glacial outburst flooding event that results from the infilling of a volcano's caldera underneath a glacier with meltwater. Once the volume of the caldera is met, the water will overflow onto the volcano's flanks. The discharge of large jökulhlaup events can be greater than 10,000 m^3/s and can be an order of magnitude bigger, to create catastrophic flooding condition (>100,000 m^3/s). The energy of this moving water from high slope to low slope can have erosive power that can erode gorges, waterfalls, and alter stream channel geometry. Carrivick et al. (2004) describes such formations resulting from Holocene jökulhlaup events along the Jökulsá á Fjöllum River, near the northern margin of Iceland's largest ice cap, Vatnajökull. Likely, a waterfall called Detifoss (65.8147, –16.3847) was eroded by such an event. The sediment created from erosive flooding will be deposited further down-gradient as the velocity of the water slows and can pile up as gravel bars or slackwater deposits.

ENERGY

7

CAN ENERGY BE HARNESSED FROM A VOLCANO?

What is geothermal energy?

At its simplest, geothermal energy is heat that can be harnessed from the Earth's interior. As a differentiated planet (Figure 1.1), the Earth is separated into parts (e.g., core, mantle, crust) under different temperatures and pressures, resulting in unique properties. The geothermal gradient explains how temperature rises from the core to the crust (Figure 1.3). At its hottest, the temperature at the center of the Earth is approximately 6000°C, roughly as hot as the surface of the Sun. However, the temperature dissipates away from the core, and in most regions of the world, the geothermal gradient is about 25°C per 1 kilometer of depth.

In areas that are plate tectonically active (Figure 1.5) or have a hot spot, harnessing heat from the Earth becomes possible at shallow depths (3.0 km or less). This is because magma radiates heat, turning groundwater into hot water or steam. Sometimes, this can be seen at the surface through hot springs, venting, fumaroles, mud pots, or geysers. The heat stored in the groundwater or steam can create geothermal energy. This can be turned into electricity or, straightforwardly, pumped into houses for radiant heat or hot water usage.

While archeological evidence reveals the use of geothermal waters for the last 10,000 years, the first experimental harnessing of steam for energy production began in 1904. An Italian aristocrat, Prince Piero Ginori Conti, used hot vapor from the ground to power a steam engine that produced enough energy to illuminate five light bulbs. By 1905, he had increased his power production to 20 kW, equivalent to running a household refrigerator today. His success prompted other inventors to consider capturing the Earth's natural energy.

Can volcanically active areas be harnessed for geothermal energy?

Volcanic regions where geothermal energy can be harnessed are commonly located near active convergent, transform, or divergent plate boundaries (Figure 1.5). These regions are producers of geothermal energy because active plate movement generates magma close to the Earth's surface, which gives off heat that can be utilized in a variety of capacities. For example, the subduction zone that outlines the Pacific Ocean (i.e., Pacific Ring of Fire) accounts for 10% of the Earth's surface and is replete with high-temperature geothermal energy fields (≤370°C) that have shallow reservoir depths (≤1.5 km). In addition, intraplate magma plumes (hot spots) also provide a geothermal heat source from the upwelling of magma from the mantle-core boundary. About 50 hot spots are known globally, and while they are found underneath continents, they are more commonly found below the thinner crust of the oceanic basin. While random in their occurrence, size, and temperature, hot spots can be integral in providing energy to countries outside of plate boundaries. Thus, geothermal resources are not equally distributed worldwide, making the production of other types of energy necessary. Additionally, while harnessing energy from volcanically active areas may be possible, it may not always be economically feasible, as it requires significant forms of technology.

Although the total world consumption of available geothermal energy is currently less than 1.0%, 88 countries have

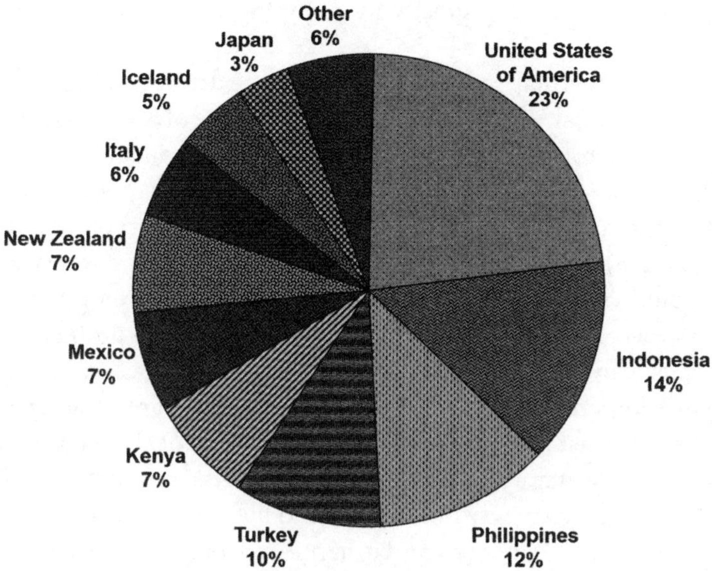

Figure 7.1 The 10 countries with the highest installed capacity (MWe) of geothermal power generation in 2020. Modified from Spijkerboer et al. (2022).

begun to harness geothermal energy (Axelsson, 2024). As of 2020, the top three producers of geothermal energy in the world were the United States (23%), Indonesia (14%), and the Philippines (12%) (Figure 7.1) (Spijkerboer et al., 2022). All three countries are located along the Pacific Ring of Fire. While the United States receives its volcanism from interplate convergence and a hot spot (i.e., Hawaii), Indonesia and the Philippines are volcanic archipelagos of many islands.

How does geothermal energy work?

Geothermal systems are categorized by geologic, geophysical, hydrologic, and engineering criteria. They consist of a permeable reservoir rock, water to carry heat from the reservoir rock to the surface, and a heat source. Most geothermal energy is either drawn directly from water naturally heated in the subsurface or indirectly by pumping water from the surface into hot rock.

Hydrothermal water can be directly extracted along most plate boundaries due to either the abundance of juvenile water exsolved from the uprising of magma or the availability of porous rock to act as an aquifer. These systems can produce hot-temperature fields (≤370°C) and have shallow reservoir depths (≤1.5 km), moderate temperature fields (≤250°C) and reservoir depths ≥1.5 km, or low-temperature fields ~190°C and depths between 1.5 and 3.0 km. For indirect extraction, hydrothermal water can be artificially produced in geothermal areas that have impermeable reservoir rock by creating large fractures. Here, at least two pumps will be placed: one of the wells will act as the water injection well; the other as the production well that brings the newly heated water to the surface for use.

Geothermal energy can be harnessed in several ways, including dry steam, flash steam, and binary cycle. Important to the United States is the technology that supports dry steam energy production. To make this energy, steam flowing underground is diverted to a turbine that drives a generator that produces electricity. The largest geothermal power plant in the world in terms of energy production, The Geysers, California (38.7450, –122.6216), uses this technology to produce 900 MW, or enough energy each year for 1 million homes from the state's northern border to San Francisco. The Geysers, which contains a complex of 18 geothermal power plants, is fueled by a magma chamber ~6.4 km below the surface. While this type of energy could also be harnessed at the famous Yellowstone Hot Spot that resides beneath Wyoming, this area is protected by National Park regulations.

Worldwide, the most common type of geothermal power plant is flash steam. Here, a pump pushes high-pressure hot water from deep inside the Earth into a tank at the surface, where it cools. As it cools, the water flashes or quickly turns to vapor; vapor drives a turbine and powers a generator. The vapor will then be retrieved in a condenser before it is placed back into the ground as water (Figure 7.2). Indonesia and the Philippines, two top geothermal energy producers, use this technology to support their economies. Lastly, a binary cycle geothermal power plant works differently than dry or flash steam. It uses two types of fluid: water and an industrial product, like tetrafluoroethane. Hot water from underground heats a second fluid, called a heat transfer fluid, in an instrument called a heat exchanger. The second fluid has a much lower boiling point than water, so it flashes into vapor at a lower temperature. When the second fluid flashes, it spins a turbine that drives a generator to produce electricity. In order to maximize their energy production, power plants in New Zealand have started adding binary cycle technology to their preexisting flash steam processes to extract energy from lower-temperature geothermal sources. The Wairakei Power Plant (–38.6180, 176.0988) is an example.

Seemingly, geothermal energy could be extracted from anywhere on the surface of the Earth. However, extracting energy is only possible if water is present to make the area

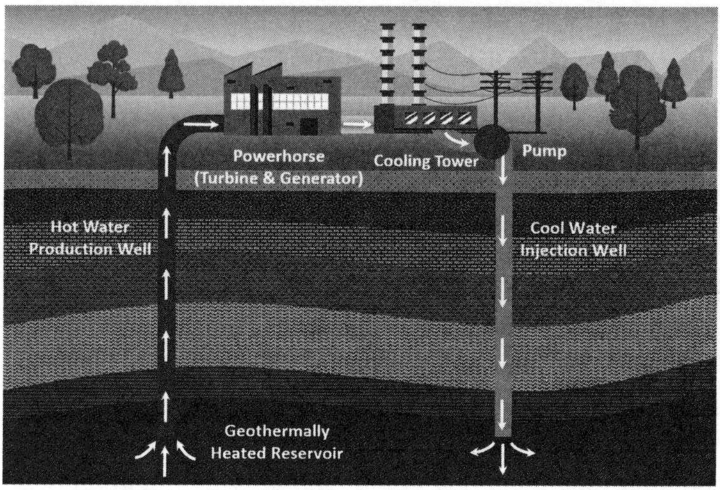

Figure 7.2 Schematic of flash steam geothermal energy production.

hydrothermal. Hydrothermal scenarios can be created indirectly through enhanced geothermal extraction (EGE). In some geothermal areas, the basement rock is impermeable and does not act as a reservoir because it does not have holding capacity. This rock is drilled in EGE to create a conduit for water to flow downward toward the heated geothermal area on the subsurface. Depending on the rock type, the depth of the well could be as deep as 5 km or as shallow as 1 km. Here, water will be forced down the well, creating new fractures acting as a reservoir. Once the water reaches boiling temperature (or greater), it is cycled to the surface through a production well to generate electricity through flash steam or a binary cycle. While this technology has successfully complemented flash steam at geothermal power plants like those at The Geysers, other countries have halted projects due to elevated seismic activity resulting from fracturing. In 2009, an EGE project was canceled in Basel, Switzerland (47.5619, 7.5883), for this reason and concern about future ground subsidence.

An innovative approach for producing geothermal energy is to tap into magmatic systems by drilling into shallow magma bodies, installing a heat exchanger, and circulating water later heated to the surface. Although this has been tested at the Hawaiian volcano, Kilauea, its shallow magma is ≤1200°C, and construction material has not yet been discovered to withstand long-term exposure to extreme temperatures and corrosivity.

What are the benefits of using geothermal energy?

The benefits of using geothermal energy overlap into three main categories: the environment, economy, and public health. The most obvious advantage of using geothermal energy for production is that it is almost a completely carbon-neutral source of emissions. Other energy, like the burning of fossil fuel, releases a gas called carbon dioxide (CO_2) into the atmosphere. In comparison, geothermal power plants emit about 35 times less carbon dioxide per kilowatt of electricity than coal power plants. The continued use of geothermal energy, although not yet everywhere in the world, can help lower the world's signature greenhouse gas emissions, which is important for reducing global temperatures. Likewise, this type of energy is unlimited as it is considered a renewable resource. Especially with closed-loop systems, water or steam used at the surface is reclaimed and recirculated back into the underground reservoir as with dry and flash steam. Because geothermal energy is renewable, after a facility is built, the overall cost of energy becomes less expensive ($0.05 to $0.10 per kWh). This energy production cost reduction can be passed on to citizens as lower heating and cooling bills. A "green energy," geothermal also has a relatively small footprint, which means that the production of energy affects a limited area (1–8 acres per megawatt) versus nuclear energy (5–10 acres per megawatt) or coal (19+ acres per megawatt). Among the most important benefits of using geothermal energy is its reliability. Unlike other carbon-neutral energies, like solar or wind, geothermal energy can be produced 24 hours a day, 365 days a year, regardless of weather conditions. Unless there is a need for repair or maintenance, geothermal electricity is available 95% of the time, supplying energy to a grid and consumers without delays or blackouts. This makes geothermal energy a great complement to most electrical grids because there are no limitations on how much base load it can contribute.

What are the unique applications of harnessing geothermal energy?

Geothermal is considered a clean energy source because it does not emit carbon dioxide (CO_2) in its production. Carbon dioxide is a greenhouse gas that traps heat in the atmosphere, causing temperatures on the Earth's surface to rise, resulting in global warming. Thus, geothermal energy is a carbon-neutral way to meet the electricity needs. It is also renewable and cost-efficient, and in places where geothermal energy is obtainable, the amount of energy produced is only limited by the size of the power generation facility.

With all of the positive benefits of geothermal energy production, it is no wonder that it is a sought-after energy source for energy-intensive manufacturing processes like aluminum smelting. Aluminum smelting is the extraction of aluminum from its oxide, alumina, to be used in cans and other everyday items. Smelting just 1 metric ton of aluminum requires approximately 14,000 kWh. For perspective, that is more energy than the average four-person American household uses in a year. For this reason, aluminum manufacturing companies have sought out countries with geothermal energy as an inexpensive option. For example, Iceland has three aluminum smelters that produce over 800,000 metric tons of aluminum annually. The aluminum industry is a major contributor to the country's economy and provides inexpensive building materials, namely corrugated aluminum, to an island void of lumber or other building materials. Likewise, moving into the new Artificial Intelligence age will require significant amounts of electricity to run data centers. Data centers worldwide consume 1%–2% of overall power, but this percentage is predicted to rise to 4% by the end of 2030. The International Energy Agency estimates that a single ChatGPT query requires 2.9 watt-hours of electricity, compared with 0.3 watt-hours for a Google search. To meet this demand, companies like Google, mindful of its carbon footprint and rising energy costs, are building geothermal power plants in Nevada to support the requirements for this new technology.

Recently, there has been a trend to access geothermal energy to meet the increased demands for food security. Greenhouses can be heated year-round by geothermal energy harnessed directly from the ground or by tapping into water or steam resources. Greenhouse heating using geothermal energy is classified into three categories—low, middle, or high temperature—which serve different growing needs. Low-temperature heat (10°C) is found nearly everywhere in the world as it is only about 10 feet underground and can be used for growing perennials, herbs, nursery stock, and some vegetables. Mid-temperature heat (60°C–150°C) utilizes heat from thermal wells and springs to grow vine vegetables, tropical plants, and fruits. Over 40 greenhouses in Washington, Oregon, and California utilize mid-temperature geothermal heating to extend their growing season into winter. High-temperature heat (300°C) can be accessed from specific geothermal fields; however, these temperatures are not necessary for growing but are used to produce energy to run air conditioning or fluorescent lighting. High-temperature heating is commonly used in Iceland for illuminating greenhouses, as the country experiences 24 hours of darkness half the year (Figure 7.3).

Interestingly, since 2003, the City of Santa Rosa, California, has combined dry steam geothermal energy production with its wastewater recycling program. The geothermal area called The Geysers (38.7450, –122.6216) has supported energy

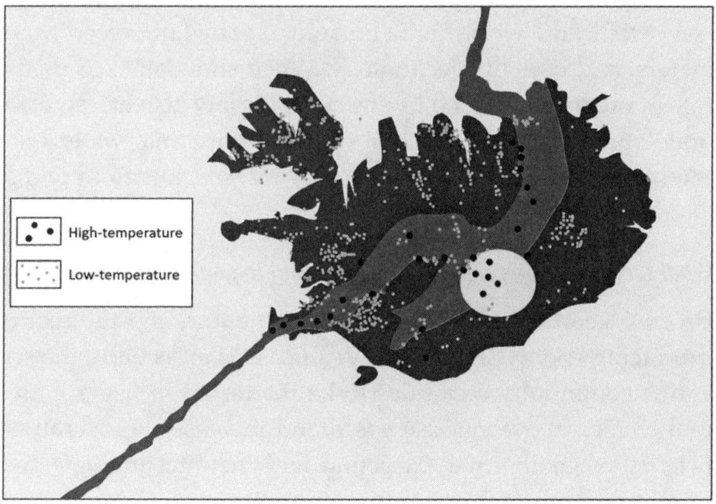

Figure 7.3 Position of high-temperature (red dots) and low-temperature (blue dots) geothermal fields in Iceland relative to the divergent plate boundary and hot spot.

demands in California since the 1970s. While The Geysers has been able to meet the needs of California's growing population over the past 50 years, the amount of steam pressure began to decrease in the underground reservoir rock from overuse. To solve this, a 42-mile pipeline was built that delivers 11 million gallons per day of tertiary treated wastewater to the high-temperature reservoir, converted to steam and directed to turbines for energy production. To date, this innovation has helped maintain The Geysers as a primary energy source for California, producing 60% of the state's energy. Perhaps most importantly, The Geysers has demonstrated that the sustainable use of geothermal energy requires a balance between the consumption and generation rate to avoid long-term impacts.

Geothermal energy has also been applied to add daily convenience to people who live in cold weather climates. Geothermal heating coils have been placed under sidewalks, driveways, roads, and bridges to keep fairways clear of snow and ice during freezing temperatures. For example, Aomori City, Japan (40.8322, 140.7469), is considered one of the snowiest cities in the world, with an annual snowfall exceeding 10 meters, and uses this technology. When snowfall exceeds the rate of melting provided by the coils, the city also has an elaborate sprinkler system that sprays hydrothermal water onto the snow to induce melting and prevent the snow from piling.

What are technological innovations relating to geothermal energy?

Because geothermal energy is carbon-neutral and the cost of producing energy is inexpensive once a plant is built, the scientific community continues to be interested in using it as a tool to reduce greenhouse gases and increase the overall efficiency of geothermal extraction with new technology. Two approaches are carbon capture and storage (CCS) and deep drilling in geothermal areas.

The process of CCS permanently stores carbon dioxide (CO_2) from the atmosphere underground in a rock reservoir with porosity, like vesicular basalt (Figure 2.6 I). But how does it work? Carbon dioxide in the atmosphere is collected and forced into water. This combination creates acidic carbonated water. This acidic solution is injected into rocks, and a chemical reaction releases cations such as calcium, magnesium, and iron into steam. Eventually, the elements will precipitate as minerals (i.e., carbonates, sulfides) within the vesicles of the rock using the CO_2 initially from the atmosphere in the process (Matter et al., 2011). This CarbFix project has been implemented using geothermal energy in Iceland at the Hellisheidi Geothermal Power Plant (64.0372, −21.4013) since 2014 to capture an estimated 2,200 tons of CO_2 annually. Iceland provides the ideal location for the CarbFix project because it has vesicular basalt as bedrock and ample water and geothermal energy to support the chemical reactions. Snaebjornsdottir et al. (2020) suggest that the active rift zone in Iceland could store over 400 billion tons of CO_2 and, if combined with the theoretical storage capacity of ocean ridges, could help offset the burning of fossil fuels.

Another important innovation in geothermal technology is drilling high-temperature wells at depths more than 5 km below the surface. These depths intersect supercritical geothermal fluids with more energy than those at shallower depths because they are at a higher temperature (450°C–600°C) and tremendous pressure (221 bar). While this makes them more efficient in generating energy, accessing them consistently and safely remains challenging for a variety of reasons:

1. The supercritical water is highly corrosive and damages expensive equipment.
2. There is a threat of magma entering into the newly drilled well.

3. Because the drill is entering a highly pressurized system, releasing volcanic gases like sulfur dioxide (SO_2) and hydrogen sulfide (H_2S) in large quantities is unavoidable.

However, with increasing knowledge and technology, more than 25 deep wells have been set in geothermal fields worldwide. These can be found in California, Hawaii, Japan, Italy, Mexico, Iceland, and Kenya. All of these wells encountered temperatures over 374°C, and in some instances, they encountered magma.

Are there any environmental impacts to using geothermal energy?

While there are many benefits to using the Earth's natural properties as energy, there are some associated environmental impacts. Like other energy sources, power manufacturing changes the landscape and can alter the area's ecosystem. Changes can influence area flora, fauna, and wildlife, and the infrastructure can be an eyesore. Unique to geothermal power plants, however, is that drilling and extraction of heat can cause an increase in microseismicity, causing low-level earthquake activity. Concurrently, this can be detrimental to air quality as volcanic gases (e.g., sulfur dioxide, hydrogen sulfide) find new pathways for release at the surface. In some places, ammonia and methane gas have also been mobilized.

The International Energy Association has suggested some guidelines as part of the Geothermal Implementing Agreement (GIA) in 1997. The GIA identified the environmental impacts of geothermal energy development to encourage its continued and sustainable extraction worldwide.

How does geothermal energy create the Blue Lagoon in Iceland?

The Blue Lagoon (63.8804, −22.4495) is one of Iceland's most visited sites—more than 700,000 tourists visit this manufactured spa annually. The 5,000 m^2 lagoon results from wastewater output from the Svartsengi geothermal plant that has operated since 1976. The geothermal water under Svartsengi is richer in minerals than typical geothermal areas due to the salted ocean water seeping through the porous lava and combining with the hot groundwater. Because the water in the Svartsengi Geothermal Power Plant has such a high mineral content (e.g., 70 ppm Cl), it has been deemed too corrosive to be used directly as a hot water source for housing. Consequently, the power plant uses the 240°C hot water to heat fresh groundwater, which in turn is pumped into homes on Reykjanes Peninsula. The characteristic milky blue color of the Blue Lagoon results from the high silica content extracted when cleaning the water before it is used in energy production. Since the mineral-rich hot water is not used directly, whatever is not used for heating is pumped directly into the lava fields surrounding the power plant, creating the Blue Lagoon. The pumped water cools during the production process, and by the time it reaches the bathing area, it is approximately 38°C. As a result of continued hot-water production at Svartsengi Power Plant, even more water is pumped into the lava fields, which means the lagoon keeps growing a few centimeters every year.

The lava fields of Svartsengi were formed in a volcanic eruption in AD 1226. Due to its young age, the lava field has a high porosity and permeability. Hydrogen sulfide that is released during the process of creating geothermal energy can corrode pipelines if not combined with ambient air. For this reason, ambient air is used to clean the system and to lower the temperatures to below 150°C, creating soft, slimy ground that lines the bottom of the lagoon. This same silica mud is used to produce skin-care products for which the Blue Lagoon is known.

Is geothermal water therapeutic?

Some geothermal areas worldwide have the great fortune of having naturally heated water flow at or near the surface to create hot springs, hot pots, or natural baths. If the temperature of these baths is between 38°C and 42°C, then they become ideal for human soaking. If they are hotter, the temperature can be reduced via cooling at the surface or by adding cold water. The mineral content of natural baths varies greatly. However, it is common for baths to contain high amounts of silica or sulfate. Both minerals are dissolution components of rocks under high temperatures and pressure, indicating that the water is coming to the surface from great depths. Sulfates, particularly magnesium sulfate or Epsom salt, are considered a particularly added benefit to geothermal pools.

It is not a coincidence that many tourist destinations are located where geothermal springs are easily accessible, as their popularity dates back centuries. Some of the earliest recordings of thermal baths are from 300 BC, with descriptions left by Greeks, Romans, and Egyptians. As an example, during the Roman Empire the popularity of natural baths rose to support 962 public baths that were frequented by citizens and soldiers. As we do today, these ancient civilizations relied on hot water for relaxation, community, and cleansing.

Some people suggest that geothermal water has healing properties. Certainly, anyone who has enjoyed a natural bath could agree that soaking in hot water relieves muscle aches and soreness relating to overuse or arthritis. Buoyancy provided by submergence in hot water can help to improve circulation by dilating blood vessels to improve blood flow and oxygenation.

Plus, there is the added benefit of stress reduction that relieves anxiety. Although research is inconclusive, many seek improvements to skin conditions or immune systems from soaking in geothermal water.

Figure 7.4 Snow monkeys enjoying a geothermal hot spring in Jigokudani, Japan. Alamay Stock Photo.

Not only do humans enjoy the benefits of natural baths, but so do Japanese macaques, also called snow monkeys (Figure 7.4). Deep within the mountains of Jigokudani, Japan (36.7327, 138.4611), a troop has been documented to soak in natural hot tubs for the past 50 years. Perhaps for relaxation, but the snow monkeys likely look to the hot water as a reprieve from harsh winter conditions (between –8.0°F and 0°F), as they rarely bathe in warmer summer months.

CLIMATE CHANGE

8

DO VOLCANIC ERUPTIONS IMPACT THE ENVIRONMENT AND CLIMATE?

How are volcanic events compared?

Because 10% of the world population lives near an active volcano, it became globally recognized in the 1980s that a scale was needed to communicate the threat of potential or ongoing eruptions within the geological community. Likewise, a way to compare historical eruptions with some accuracy was also needed so that hazard mitigation maps could be designed, especially in areas with long repose times between volcanic events. As a result, two models are commonly used to rank volcanic eruptions: the Volcanic Explosivity Index (VEI) and dense-rock equivalent (DRE).

Newhall and Self (1982) proposed the VEI as a means to categorize recent and paleo events to make comparisons based on an eruption's explosive character, using a logarithmic scale of 0–8 (lowest to highest) to depict magnitude. The criteria included in the VEI are listed in order of decreasing reliability: volume of ejecta, column height, qualitative description of the event (e.g., gentle, explosive, colossal), classification (e.g., Hawaiian type, Vulcanian type), duration (hours of the continual blast), tropospheric injection, and stratospheric injection. For example, a value of 0 can represent a nonexplosive event that produces $<10{,}000\ m^3$ of tephra, and a value of 8 can represent a cataclysmic event that produces $1.0 \times 10^{12}\ m^3$ tephra (Figure 8.1). Therefore, the VEI for modern or ancient

CRITERIA	VEI 0	VEI 1	VEI 2	VEI 3	VEI 4	VEI 5	VEI 6	VEI 7	VEI 8+
Description	Non-explosive	Small	Moderate	Moderate large	Large	Very large →	→	→	→
Volume of ejecta (m^3)	$<10^4$	10^4-10^6	10^6-10^7	10^7-10^8	10^8-10^9	10^9-10^{10}	10^{10}-10^{11}	10^{11}-10^{12}	$>10^{12}$
Column height (Km)	<0.1	0.1-1	1-5	3-15	10-25	>25 →	→	→	→
Qualitative description	← " Gentle, effusive" →		← " Explosive" →		← " Catalysmic, paroxysmal, colossal" →				
				" Severe, violent, terrific" →					
Classification		← " Strombolian" →			" Plinian" →				
	← "Hawaiian" →		← "Vulcanian" →			"Ultraplinian" →			
Duration of continuous blast (hours)	← <1 →				← >12 →				
			← 1-6 →						
				← 6-12 →					
Tropospheric injection	Negligible	Minor	Moderate	Substantial →	→	→	→	→	→
Stratospheric injection	None	None	None	Possible	Definite	Significant →	→	→	→
Examples	Kīlauea	Nyiragongo (2002)	Mount Sinabung (2010)	Nevado del Ruiz (1985), Soufrière Hills (1995)	Mount Pelée (1902), Eyjafjallajökull (2010)	Mount St. Helens (1980)	Krakatoa (1883), Mount Pinatubo (1991)	Tambora (1815)	Yellowstone (640,000 BP) Toba (74,000 BP)

Figure 8.1 Comparison of variables to categorizing eruption styles representing the Volcanic Explosivity Index. Modified from Newhall and Self (1982).

events can be approximated by eruption magnitude (volume) or intensity (mass eruption rate).

Newhall and Self (1982) acknowledged that there are limitations to the VEI. In particular the scale does not include atmospheric or geographic (i.e., latitude proximity) data, nor does it attempt to classify injection type. Additionally, there are some challenges in applying the VEI to phreatic events, which sometimes only produce small eruptive volumes of tephra that are locally dispersed, and to effusive eruptions, where column height and ejecta are not applicable.

An alternative way to compare volcanic eruptions is by identifying the dense-rock equivalent (DRE) of ejected material. This method estimates the total volume of magma that erupted during a volcanic event through a mathematical calculation correcting for void spaces in specific substrates that have porosity. This correction can be made by comparing the bulk density of the tephra deposit with the known density of the rock type that makes up the tephra. Like the VEI, this method can be applied to either explosive eruptions that eject tephra or nonexplosive eruptions that effuse lava. The DRE is thought to be particularly useful in approximating the volume of ancient eruptions because data can be obtained by mapping the distribution and thickness of tephra deposits preserved within the geologic rock record. However, processes of compaction and erosion have to be considered. An example of using DRE to understand the scope of a historical eruption was applied to the 1875 event at Askja, Iceland (1,371 m; 65.0351, −16.7427) (Carey et al., 2010). Here, the authors correlate historical eyewitness accounts of column height and eruptive time used to establish a VEI of 4.0 to a tephra layer used for DRE (Figure 8.2) that was found both on the island of Iceland and more than 1,000 km across the North Atlantic in the Scandinavian geologic record (Figure 8.3).

Thus, the VEI and DRE can be used as complementary models and are best used as a check and balance using

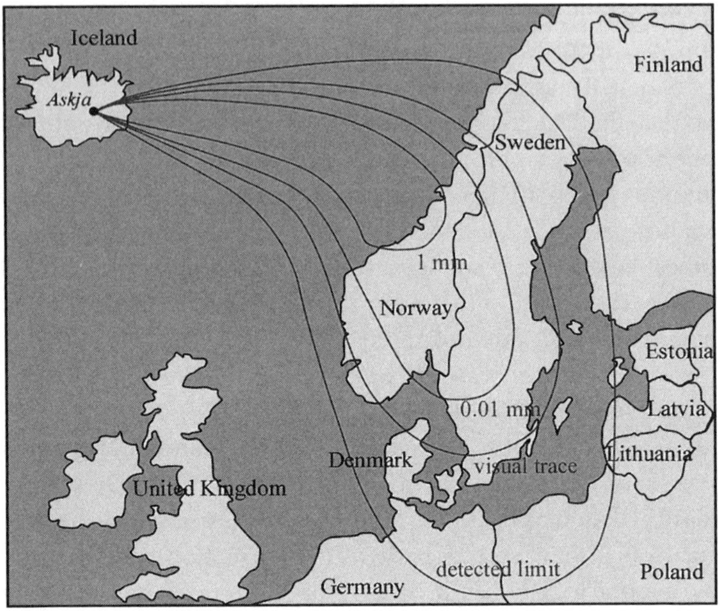

Figure 8.2 Askja 1875 ash fall dispersion map with associated thickness of deposition. Modified from Carey et al. (2010).

Event	Deposit	Column Height (km)	DRE (km^3)	Tephra Volume (km^3)	Eruptive Time (hours)
Askja 1875 Unit B	Subplinian	8	0.004	0.014	1
Askja 1875 Unit C	Phreatoplinian	22.8	0.104	0.45	1
Askja 1875 Unit D	Plinian	26	0.213	1.37	6

Figure 8.3 Askja 1875 column height, dense-rock equivalent (DRE), and tephra volume as compared to eruptive hours. Unit B, C, D refers to eruption episodes. Modified from Carey et al. (2010).

eyewitness accounts (when possible), satellite imagery, and monitoring instruments, followed by field surveys. Newhall and Self (1982) estimated VEIs for some 8,000 eruptions globally and compiled a list of the largest explosive eruptions in the world since AD 1500 with VEI ≥ 4. Some of those are found in Figure 8.4 with their correlating DRE. This figure includes

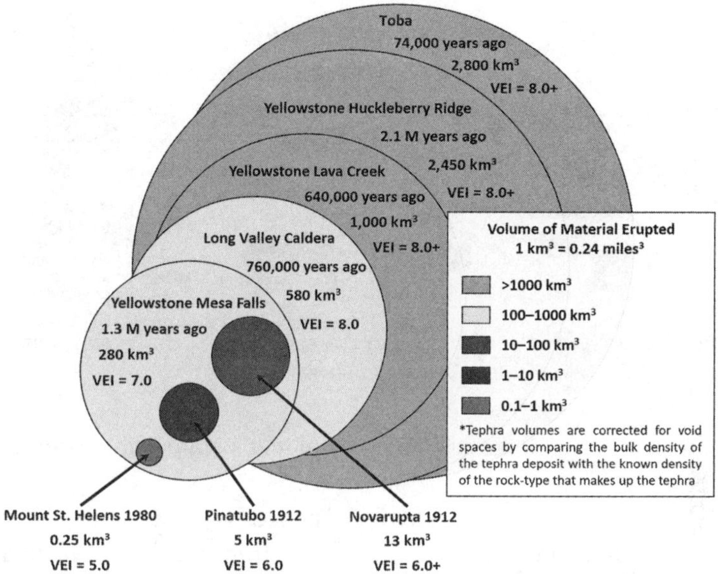

Figure 8.4 Comparison of eruption volumes in dense-rock equivalent (DRE) and Volcanic Explosivity Index values for several events. Modified from the United States Geological Survey.

information for Toba, Northern Sumatra, Indonesia (2.8191, 98.6077; 1,597 m). The Toba supervolcano is considered to be the largest eruption in the last 2 million years, releasing an energy equivalent of about one gigaton of TNT to produce 2,800 km^3 DRE.

Can volcanic eruptions change the climate?

Evidence indicates that volcanic eruptions can cause the global climate to change, making the Earth's surface hotter or colder. However, the duration of temperature flux depends on *how* the atmosphere (or higher) was affected (i.e., volcanic ash, gases) and by *what* mechanism (i.e., explosive or nonexplosive eruption style). While the length of disturbance can be at any resolution from days to years to centuries and beyond, the consequences of heating or cooling can have significantly

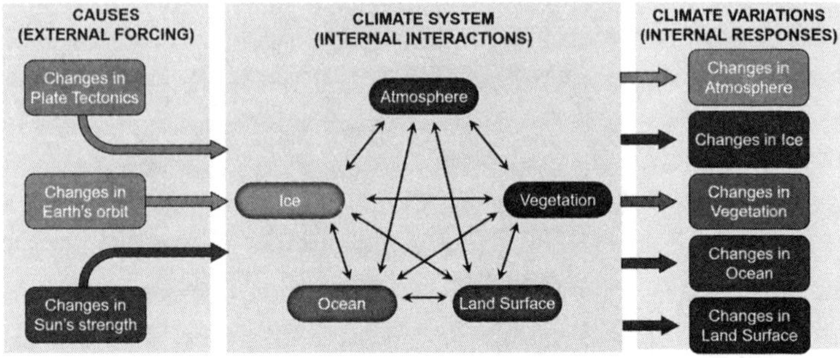

Figure 8.5 A depiction of how Earth's external forcings can alter internal interactions causing system responses. Modified from Ruddiman (2014).

longer-lasting impacts on the planet's internal response (Figure 8.5).

To understand how climate influences volcanism, one must first understand how climate differs from weather. Weather results from temporary conditions, like rain, wind, or cloud formation, occurring within the atmosphere (Figure 8.6). The atmosphere, called the troposphere, is the distance from the ground surface to ~10 km above. Beyond the troposphere is a 10-km-wide band called the tropopause that acts as the boundary before entering into the stratosphere, which starts at ~21 km above the ground surface and ends at approximately 50 km. Weather does not occur in the stratosphere as it contains very little moisture and significantly less turbulent conditions due to a drastic reduction in pressure. Climate refers to long-term patterns, usually from years to decades, to reflect average weather conditions in a region describing tropospheric conditions like temperature, precipitation, winds, humidity, and seasonal cycles.

Thus, volcanism can influence short-term weather patterns and long-term climate depending on the eruption's intensity, as defined by the VEI (Figure 8.1). The most important considerations will be the volume of ejecta, column height, duration of continuous blast, and whether the particulate matter has exceeded the troposphere. If the column height is less than 20 km,

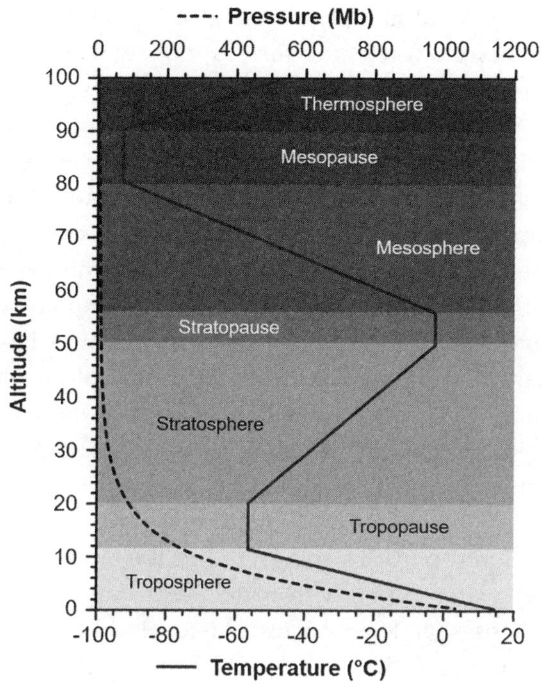

Figure 8.6 A comparison of temperature and pressure to increases in altitude above the Earth's surface. Modified from coolgeography.co.uk.

the particulate matter does not leave the troposphere. This means that in whatever quantity emitted, ash will be carried by weather patterns established by the volcano's latitudinal position on the globe. Although the size of ash particles can vary, most will be too heavy to be carried great distances, and most will fall out near the vent. During the 1980 Mount St. Helens, Washington (2,459 m; 46.1919, −122.1955), eruption, the column height grew to ~19 km in less than 15 minutes. Within hours, 13 mm of ash fell on Spokane, Washington (~366 miles away), and the city went into darkness. The temperature fell 8°C because incoming solar radiation from the sun was blocked. The cold and darkness were temporary and conditions returned to normal the following afternoon when the plume dissipated.

Furthermore, although ash traveled the globe within the troposphere for a week after the eruption, global temperature was unaffected. This is because the troposphere has a maximum residence time of 10 days. It holds moisture and associated particulate matter within this reservoir for under 2 weeks before it is recirculated in the water cycle via precipitation.

The amount and type of gases and aerosols emitted from a volcanic eruption can lead to either global warming or cooling. Here, defining the difference between a gas and an aerosol is essential. A gas is a state of matter that does not have a defined shape or volume. Gases emitted from volcanoes primarily include water vapor, which can represent more than 80% of the volumetric release, but can also include some amounts of carbon dioxide (CO^2), carbon monoxide (CO^-), and sulfur dioxide (SO^{2-}). Aerosols are a mixture of very small solid (i.e., ash) and liquid particles suspended in the air. Consequently, mixing of water vapor with some gases can lead to chemical reactions that form harmful aerosols like sulfuric acid (H_2SO_4), hydrogen fluoride (HF^-), hydrogen chloride (HCl^-), or hydrogen bromide (HBr^-), which can be carried around the world if they enter the stratosphere.

Three example eruptions will be used to illustrate the impact of gases and aerosols on global climate: Hunga Tonga-Hunga Ha'apai (89 m; −20.5358, −175.3819); Laki, Iceland (651 m; 64.0752, −18.2405); and Mount Pinatubo, Philippines (974 m; 15.1444, 120.3334). The eruption of Hunga Tonga occurred on January 15, 2022, in the South Pacific Ocean. The underwater explosion produced the tallest instrumentally recorded eruption column, which exceeded the upper limits of the stratosphere at 58 km above the Earth's surface (Figure 8.6). Accordingly, the National Oceanic and Atmospheric Administration (NOAA) reported that 150 million tons of water vapor immediately entered the stratosphere (NOAA, 2023). NOAA models indicate that global climate could warm by 1.5°C over the next 5 years, resulting from the unprecedented amount of water vapor, a greenhouse gas, entering the system. Oppositely, eruptions at Laki (Iceland) and Mount Pinatubo (Philippines) led to global cooling. A historically well-documented fissure eruption at Laki lasted 9 months, from June 1783 to February 1784. While a basaltic eruption, the mixing of glacial meltwater with magma created a phreatic eruption style that pushed large amounts of gas and ash 9–13 km into the troposphere and lower stratosphere. Thordarson and Self (2003) estimate that the amount of H_2SO_4 aerosol produced from this eruption was about 200 Mt (megatons). When an abundance of H_2SO_4 is present, it acts as a shield from incoming solar radiation, causing the planet to cool. In addition to acid rain and acid fog that veiled northern Europe for 5 months, the sulfuric aerosol stayed in the stratosphere for a year, causing Northern Hemisphere temperatures to fall by 1.5°C for the next 3 years, which could have triggered phases of famine that are well documented in Europe during this time. Similarly, the eruption at Mount Pinatubo on June 15, 1991, injected 20 million tons of SO_2 gases and 10 km^3 ash into the stratosphere, causing the global climate to decrease by 0.5°C for 2 years (1991–1993) (Fiocco et al., 1996).

Are flood basalts linked to mass extinction?

"Flood basalt" refers to the building of expansive lava plateaus covering areas of more than 1 million km^2 that formed from individual flows with volumes up to thousands of km^3. Usually, the lava flows erupt over a total time interval of less than 1 MY. A result of large-scale effusive volcanism is the emittance of anomalous greenhouse gases into the atmosphere that can force abrupt cooling or warming. Moreover, because of the abrupt temperature change, most plant and animal species do not have adequate time to adapt or evolve to new climatic conditions, which can lead to their extinction. In geology, *mass extinction* is defined as an event that occurs when at least 75% of all living populations die within a confined period of time that can be thousands to millions of years.

Over the past 250 MY, there have been several episodes where large outpourings of effusive basaltic lava have been correlated with extinction and distinct faunal changes confirmed by the stratigraphic record (Figures 8.7 and 8.8). Although the correlation between some flood basalt episodes

Flood basalt episode	Age (Mya)	Volume (10^6 km^3)	Duration (Mya)	Stratigraphic boundary
Columbia River	16 ± 1	0.25	≤1 (for 90%)	Early/Mid-Miocene
Ethiopian	31 ± 1	≤1.0	≤1	Early/Late Oligocene
North Atlantic	57 ± 1	>1.0	≤1	Paleocene/Eocene (Thanetian/Selandian)
Deccan	66 ± 1	>2.0	≤1	Cretaceous/Tertiary
Madagascar	88 ± 1	?	≤6?	Cenomanian/Turonian (Turonian/Coniacian)
Rajmahal	116 ± 1	?	≤2	Aptian/Albian
Serra Geral/Etendeka	132 ± 1	>1.0	≤1 or ~5?	Jurassic/Cretaceous (Hauterivian/Valanginian)
Antarctica	176 ± 1 or 183 ± 1	>0.5	≤1?	(Aalenian/Bajocian)
Karoo	183 ± 1	>2.0	0.5-1	Early/Middle Jurassic
Newark	201 ± 1	>1.0	~0.6	Triassic/Jurassic
Siberian	249 ± 1	>2.0	≤1	Permian/Triassic

Figure 8.7 Flood basalt episodes of the last 250 million years and associated faunal events identified in the stratigraphic record. Modified from Rampino and Self (2000).

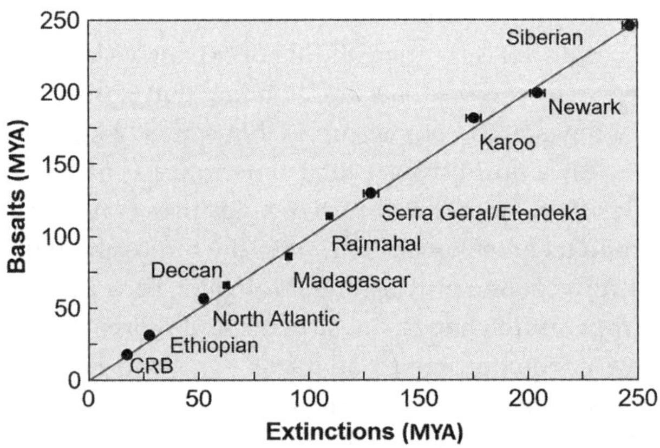

Figure 8.8 Possible correlation of flood basalts and extinction/faunal changes. Location names correlate to Figure 8.7. Modified from Rampino and Self (2000).

and extinction is accepted chiefly within the scientific community (e.g., Siberian flood basalts causing the Permian-Triassic Mass Extinction), other flooding episodes may only be concluded as one variable contributing to a changing planet (e.g., Deccan flood basalts and the Cretaceous-Tertiary Mass Extinction killing the dinosaurs). Moreover, because flood basalts are directly linked to intense mantle plume activity, a host of other geological factors could coincide and contribute to a changing climate (e.g., rifting events, changes in seafloor spreading rates, and increased tectonism).

Among the most extensive flood basalt deposits in the world is the Siberian Lava Traps, Russia (61.6755, 98.8677). Approximately 250 MYA, a mantle plume originated under Siberia and produced enormous amounts of lava for 1 MY at a rate of 1.3 km^3/year (Renne and Basu, 1991). The total flow covers an area of ~5×10^6 km^2 with a volume of ~ 4×10^6 km^3 (Ivanov et al., 2013), equivalent to covering the entire United States with 1 km of lava. Wu et al. (2021) estimate that the cumulative release of greenhouse gases caused global levels to increase sixfold from start to finish, creating an amplifying feedback loop of warming that likely initiated the Permian-Triassic Mass Extinction (PTME) as temperatures ramped 15°C or more (Bond and Grabsy, 2017). This sudden and drastic increase in temperature influenced all climatic systems (Figure 8.5), resulting in the death of vegetation, reduction in photosynthesis, warming and acidification of the oceans, reduction in carbon sequestration, and more (Saunders and Reichow, 2009). The PTME, also referred to as the "Great Dying," is considered to be the worst mass extinction in geologic time, in which 96% of all marine species and 70% of land species perished.

How far can volcanic ash travel?

Volcanic ash is a mixture of rock, mineral, and glass particles ejected from a volcano during a volcanic eruption. The distance that volcanic ash can travel will depend on many factors, allowing it to fall locally or be carried thousands of miles away from its source. The first thing to consider is the amount of ash available for transport. Although any volcanic eruption can produce ash in some quantity, those volcanoes that are abundant producers have a silica-rich (felsic) magma source (Figures 2.3 and 2.4). This is because the highly viscous felsic magma generates explosive eruptions that will shatter the magma into tiny fragments as it is propelled into the atmosphere, where it will solidify. Furthermore, the volatility of an eruption will be intensified if water is present, creating a hydromagmatic or phreatic eruption style. As a rule, the longer a volcano has been dormant, the larger the volcanic eruption will be in terms of volume of ejecta, column height, duration, and the energy to force gases and ash into the stratosphere (Figure 8.1). Volcanic events classified as Vulcanian, Pilian, or Ultra-Pilian are large ash producers, with a VEI equal to 3.0 or higher. These large-scale events can also be followed by caldera collapse, generating more suspended material.

The location and timing of volcanic events will also affect the distribution of volcanic ash. Thus, wind direction and strength in the upper atmosphere are significant variables that change seasonally so that ash from an event could travel a lesser or greater distance. The impact of wind direction can be illustrated using Poas, Costa Rica (2,497 m; 10.2097, –84.2305), as an example (Figure 8.9). In this diagram, note that the distance the ash traveled in January 1994 (<3 km) is considerably shorter than that in July–August 1994 (>40 km). While winds always blow from east to northeast throughout the year, they are usually less than 15 km/h. However, in Costa Rica, in July, more vigorous trade winds (called the Pacific vernaillo) bring dry weather during the rainy season and increase the wind speed to more than 30 km/h. Thus, the distribution of ash, as well as the number of communities affected, was greater in January with higher wind speeds.

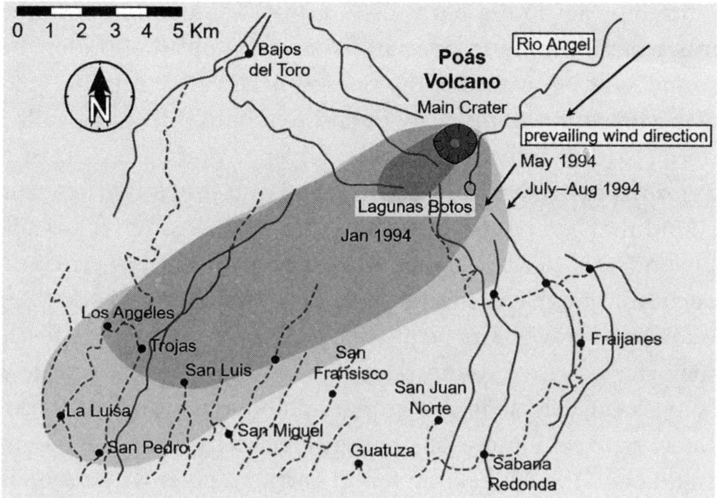

Figure 8.9 Poas wind direction for 1994 eruptive episodes. Modified from Mora Amador et al. (2019).

Another variable to consider in the distribution of volcanic ash is its grain size. The size and shape of the ash are determined by those of the gas bubbles in the magma before eruption. The solidified particle size within ash clouds is considered fine-grain and usually less than 2.0 mm in diameter. Smaller grain sizes will have less density and can be transported farther. Whether they are transported in the troposphere or the stratosphere depends on the maximum column height of the ash cloud during peak eruption (Figure 8.1). Ash carried in the stratosphere usually has a diameter <0.5 mm, and while it can be spread across the globe, most of it will settle back to the surface within three months post-eruption.

How loud is a volcanic eruption?

Sound is measured by an instrument called the audiometer, which was invented by Alexander Graham Bell in 1879. To create sound, a pressure wave is generated by something vibrating that forces particles to move. The more intense the movement, the more pressure is created, and the louder the sound will be; consider the energy it takes for a polite golf clap versus loud applause. Sound is reported in units called decibels (dB). For comparison, a whisper is about 20 dB, and a normal conversation is 60 dB. The loudest possible acoustic sound that can be measured in the air will register at 194 dB. Beyond this decibel, sound waves become altered, create a vacuum, and have enough energy to transform themselves into shock waves. Consequently, shock waves, moving faster than the speed of sound (1,234 km/h at 20 °C), can create a sonic boom. Thus, decibels reported above 194 dB more accurately represent immense pressure created by force that moves so quickly it will break the sound barrier. This is why sound is also reported in units of sound pressure level (SPL).

Volcanic eruptions create the loudest sounds on the planet. It has been estimated that the loudest sound on Earth resulted from the Plinian-style eruption (Figure 8.1) of Krakatoa on May 20, 1883. At its source, the eruption produced a sonic boom equivalent to 310 dB and sent shock waves around the globe four times, altering atmospheric pressure. This eruption has been classified as a VEI 6 and is thought to have released as much energy as 200 megatons of dynamite (TNT). For reference, the atomic bomb dropped on Hiroshima, Japan, in 1945 to end World War II released the equivalent of 0.015 megatons of TNT. More recently, the Hunga Tonga-Hunga Ha'apai (89 m; −20.5358, −175.3819) event on January 15, 2022, produced similar results. Audiometers almost 10,000 km away in Anchorage, Alaska, recorded sounds at 140 dB, arriving 8 hours after the VEI 6 eruption. Eyewitness accounts compare the sound of the boom to that of a shotgun being fired.

Why are lightning storms sometimes associated with volcanic eruptions?

There is yet to be a concise formula for determining whether lightning will form during a volcanic eruption, as many variables are outlined in the VEI (Figure 8.1). However, lightning associated with volcanism typically forms under scenarios where volcanic ash is present as well as an abundance of water in either solid, liquid, or gaseous form.

Lightning more commonly occurs with explosive eruptions that produce an ash column between 7 to 12 km and 1 to 4 km in height. The former elevations correlate to a region in the troposphere where thunderstorms are also formed. Here, moisture-rich convection currents influence eruption plume dynamics that can trigger volcanic lightning. The latter elevations correlate to electrical behavior close to the vent as energy interacts with the ground surface (McNutt and Williams, 2010).

Many recently observed volcanic eruptions have generated lightning. The following list demonstrates the wide variability in explosivity that ranges in intensity from a VEI of 6 to 0, indicating that innate properties of the ash column influence the generation of electricity: Eyjafjallajökull, Iceland (2010); Etna, Italy (2013, 2015, 2021); Sinabung, Indonesia (2014); Villarrica, Chile (2015); Colima, Mexico (2015, 2016, 2017); Ambae, Vanuatu (2018); Anak Krakatau, Indonesia (2019); Taal, Philippines (2020); Sakurajima, Japan (2021); St. Vincent, Caribbean (2021); and Hunga Tonga-Hunga Ha'apai, Indonesia (2022). Some volcanic eruptions can produce innumerable lightning strikes; almost 200,000 bolts were recorded during the 2022 Hunga Tonga- Ha'apai event.

When lightning strikes the ground surface, a geological formation called a fulgurite can be created. Essentially, when a bolt comes into contact with sand, soil, or other substrates, they will metamorphose, forming glassy tubes. The heat generated by volcanic lightning (up to 30,000°C) in eruptive columns and plumes can cause the ash particles to melt and solidify, forming exceptionally small glass spherules (~50 μm) intermixed with the ash fall. These have been documented at Mount Redoubt, Alaska (2009), and Eyjafallajökull, Iceland (2010) (Genareau et al., 2015).

The first written account of lightning associated with volcanism was provided by Pliny the Younger, who observed the devastating eruption of Mount Vesuvius, Italy, in AD 79. His well-preserved letters provide insight as he writes, "There was a most intense darkness rendered more appalling by the fitful gleam of torches at intervals obscured by the transient blaze of lightning." His letters also reveal that the volcanic event killed his father, Pliny the Elder. The Plinian-style classification was created as a recognition of the important scientific documentation provided by the letters of Pliny the Younger.

How quickly can life rebound after a volcanic eruption?

Whether mafic or felsic in composition, over time, all volcanic areas become fertile environments that provide fabulous habitats that can establish unique ecosystems for biota to thrive. The question of how quickly life can rebound after a volcanic eruption will ultimately depend on the eruption style, as different types of damage can result from the effects of volcanic gases, explosive blasts, pyroclastic flows, lahars, lava flows, or tephra (ash) fall (Figure 5.1). Furthermore, the extent of the damage differs with the chemical and physical properties of the eruptive products, the time of year, the amount of snow and ice cover, the behavior of organisms, the type of vegetation, and even the time of day of the eruption. Because each volcanic event is unique, and because there are so many variables involved in describing such an event, it is often difficult to predict the rate of return of life. However, interesting data has been collected post-eruption that can be compared and analyzed to understand ecosystem recovery.

The rate at which life rebounds post-eruption directly correlates to the extent of damage. That is, the greater the devastation, the longer it will take for an ecosystem to repair and rebuild. Recovery time is best illustrated by hazard mitigation maps that identify zones from high to low impact (Figure 8.10). The worst-case scenarios are found within the blast zone nearest the volcanic vent. This is where not only is all vegetation stripped but also the soil has been vaporized, leaving

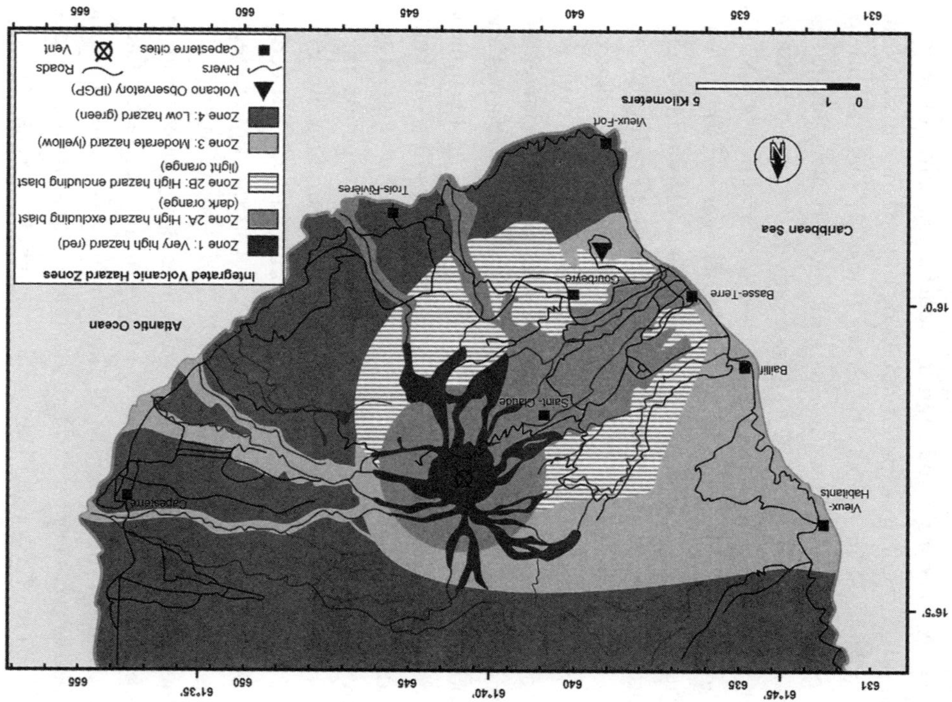

Figure 8.10 An example of a volcano hazard zone map completed for Guadeloupe (16.2875, –61.5503) located in the Lesser Antilles. Modified from the University of West Indies Seismic Research Center.

behind barren, exposed bedrock. For example, several years after the May 18, 1980, eruption of Mount St. Helens, the blast zone appeared much as it did during the immediate aftermath. This is because the development of soil is a lengthy process, and without it, the colonization of vascular plants cannot take place. Areas considered high to moderate hazard are often in the direction of anticipated lahars or pyroclastic flows. While both lahars and pyroclastic flows transport material downslope, pyroclastic flows are usually more destructive to the environment because they are not only fast-moving but also carry a hot mixture of ash, gas, and rock fragments. On the other hand, lahars are moving blocks of vegetated soil down a gradient and sometimes can remain intact. Herein lies an important distinction between complete habitat eradication and species preservation. When blasts or pyroclastic flows occur, they can completely erase all signatures of life. However, with lahars, vegetation often lives but is transported to lower elevation areas, which can cause shifts in ecological niches and landscape change. Thus, with the survival of life post-eruption, species revitalization can occur much more rapidly than if new colonization needs to rise. In terms of ashfall, dispersal will often correlate to all zones on the hazard mitigation maps, with the deepest deposits occurring closest to the vent (Zone 1) and thinning in the direction of dominant wind direction. However, ash falling near steep vents can quickly erode, as was documented at Irazu, Costa Rica (3,223 m; 9.980, −83.8491), when one-third of the total volume was removed from its slopes within one year after the 1964 eruption cycle. On flatter landscapes, the thickness of the ash has been linked to the speed of recovery for farmsteads in Iceland; farmland covered in <10 cm of ash will likely be abandoned for a year, 15–20 cm are abandoned for 1–5 years, and settlements covered in 20–50 cm of ash are abandoned for decades. Likewise, historical accounts provide evidence of coffee plantations buried in 1 m of ash rebounding 1 year after the 1902 eruption of Santa Maria, Guatemala (3,651 m; 14.7569, −91.5533).

The establishment of life on brand-new oceanic islands is the best model for identifying ecosystem succession. The volcanic island of Surtsey, about 15 km off of the southern coast of the island, was among the first to be thoroughly documented by geologists and evolutionary biologists from its first explosion on November 14, 1963, to the present day. As soon as the island stabilized above sea level in 1965, the first evidence of vascular plants was found, which were followed by bryophytes (1967) and lichen (1970). Additionally, sea birds (mainly gulls) started to nest on the island in 1970, and became an integral part of seed dispersal in the island interior. Today, there are more than 69 species of vascular plants. This island remains untouched by humans, as it remains a pristine natural laboratory and is protected as a United Nations Educational, Scientific, and Cultural Organization (UNESCO) World Heritage site.

How do animals respond to volcanic eruptions?

Like humans, animal habitats and populations can be severely impacted during a volcanic event. If there is a precursory seismic warning, then some large mammals have been reported to migrate away from the epicenter to safer areas where shaking cannot be felt. Likewise, animals may also be spared from the effects of lava flows or lahars by simply moving out of the way or finding refuge on a kipuka. Small burrowing animals can typically withstand an eruption and, along with birds, are among the first to emerge from the devastation.

Does climate change increase volcanism?

Climate change can increase volcanism in particular landscape scenarios where melting glaciers and volcanoes are present. Glacial ice exerts considerable weight (917 kg/m^3) on a surface. When a glacier melts, the weight on the bedrock is reduced, and the bedrock begins to uplift as pressure becomes less. This process is referred to as isostatic rebound. In this scenario, decompression melting can occur that causes the mantle to melt and form magma, which can lead to volcanic eruptions. In areas like Iceland, which has both glaciers and a hot spot, volcanism has been shown to increase as glaciers thin, allowing fractures to form in the bedrock that can act as conduits for magma to approach the surface more easily. While this can be witnessed today, the island experienced significant uplift with rates around 6.9 cm/yr during the Holocene, which has been linked to a distinct pulse of volcanism.

PREDICTION

9

CAN VOLCANIC ERUPTIONS BE PREDICTED OR PREVENTED?

How many active volcanoes are there in the world today?

An active volcano, as defined by the Global Volcanism Program (GVP), is one that has had an eruption within the past 10,000 years, a period known as the Holocene Epoch. This time frame is significant as it provides a geological context for the activity of a volcano, allowing us to understand its recent history and potential for future eruptions. The GVP, a database maintained by the Smithsonian Institution National Museum of Natural History, was first launched in 1968 and remains an ongoing log of the world's active volcanoes. GVP monitors 1,500 active volcanoes and has isotopically dated ash from ~8,000 eruptions. The monitoring network indicates that 50–70 volcanoes are erupting on the planet at any given time (Global Volcanism Program, 2024). However, the actual number of active volcanoes on the planet is much greater, as there are thousands of underwater seamounts whose eruptions go unrecognized and unrecorded (Figures 1.7 and 1.8). Figure 9.1 illustrates the distribution of volcanoes by nation in terms of their activity since the Holocene and the 20th century. While it may be foreseeable that Indonesia leads the world in the number of historically active volcanoes (e.g., Tambora, 1815; Krakatoa, 1883), it may be surprising that the United States holds a prominent position, with eruptions in Hawaii, Alaska, and Cascades, Marianas, and America Samoa (Figure 3.5).

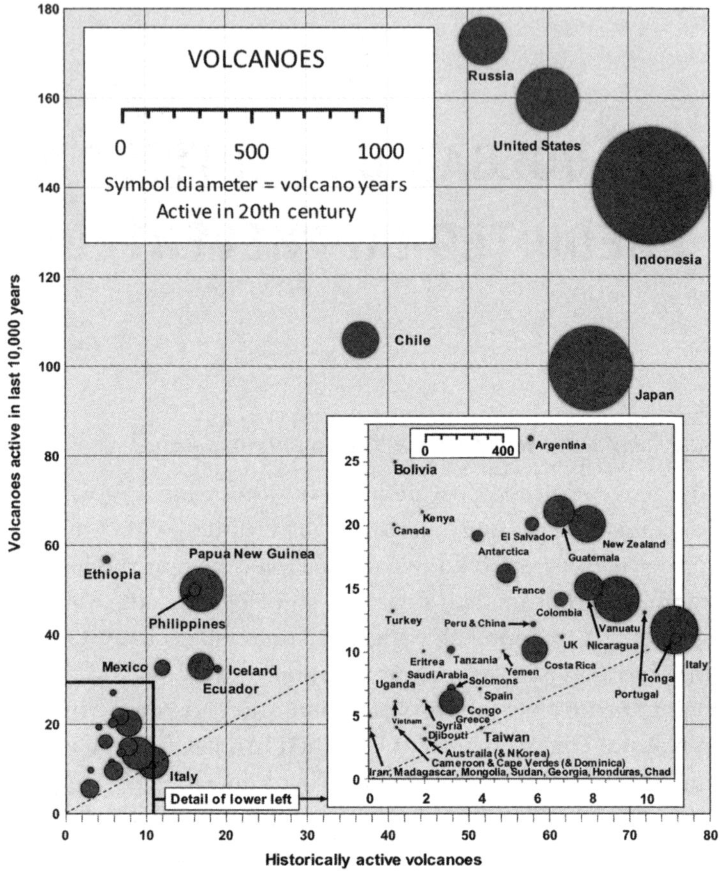

Figure 9.1 Volcano distribution by nation. The 56 nations shown here contain 98% of the world's historically active volcanoes, and another 25 contain at least one volcano active in the last 10,000 years. Distance from the dashed line may indicate a short historical record, relatively thorough study of young-but-not-historical volcanoes, or a combination of the two. Circle diameter shows the wide range in volcanism within this century. Modified from Simkin and Siebert (2000).

How often do volcanoes erupt?

Frequency measures how often an event (volcanic eruption, earthquake, flooding, etc.) will occur and is vital to understanding natural disaster risk. Usually, the more frequently an event occurs, the lesser energy (intensity) the system will have, and therefore, it will be less destructive. Here, we can compare a casual rainy day to a flooding event, as rainy days are less threatening and happen more often. The practical application of this understanding is seen in a statistical formula called the recurrence interval (RI), which calculates the average time between eruptions. This is not just a theoretical concept, but a practical tool that can be used to estimate the risk of a volcanic event.

The equation to solve for RI is listed below:

Eq. 1

$$\mathrm{RI} = n(\mathrm{years}) + 1/M$$

where
RI = recurrence interval, n = number of years of data, and M = magnitude of the event.

The RI is solved by acquiring the number of times a volcano has erupted in the past (n) and correlating that event with its intensity (M). For volcanoes, the event's magnitude is described by a logarithmic scale called the Volcanic Explosivity Index (VEI). The VEI measures the volume of tephra (ash, lava fragments, and rock particles) ejected during an eruption, the height of the eruption column, and the duration of the eruption. Inversely applied to the RI, the most significant event in the data set will be a magnitude of 1. Thus, if you have 99 years of data (n + 1, or 99 + 1) and you divide by the most significant event recorded for the last 100 years (M = 1), you end up with a recurrence interval of 100 years. For this data set, the largest volcanic eruption is expected to happen once every

100 years. RI is often compared to another statistical model called probability (P). Probability is used to estimate the percentage, or likelihood, of an event to occur each year. Often, insurance adjusters make maps that include probability to determine the risk of a natural disaster occurring to a property. To solve for P, 1 is divided by RI and multiplied by 100%, as seen below:

Eq. 2

$$P = 1/R1 * 100\%$$

where
P = probability (%) and RI = recurrence interval.

For the aforementioned data set, 1 divided by 100 years and multiplied by 100% would total 1%. This means that there is a 1% chance *every year* that a magnitude 1 volcanic event could occur. Based on probability, larger-intensity events are less likely to occur more frequently than those of lesser magnitude or energy.

Here are two real-world examples of RI and P, for the Yellowstone Caldera, Wyoming (2,584 m; 44.4127, −110.7230), and Mauna Loa, Hawaii (3,986 m; 19.4727, −155.5919). From ash deposits found in the geologic record, it is understood that there have been three major eruptions at Yellowstone over the last 2.1 MY. This results in a large eruption occurring about every 700,000 years (RI). This results in an extremely small chance ($P < 0.0000001\%$) of an eruption of equal size occurring each year. More recently, data from Mauna Loa has been analyzed for RI and P. In the past 167 years, Mauna Loa has erupted 33 times, resulting in an eruption every 5 years (RI). This means that in any given year, there is a 20% chance that an eruption will occur.

While RI and P are commonly used as tools to comment on a volcano's eruption cycle, it's important to be aware of their limitations. These statistical models are not considered predictive, as they are often unreliable. First, the RI is only as good as the eruption data set available. The value (n) is considered a limiting factor, meaning that it significantly restricts, or limits, the result of the calculation. Thus, the less data available, the less likely the RI will be accurate. For volcanoes, it is possible to extend eruption data sets by looking for ash (tephra) in geologic records. This is the study of tephrochronology. Using isotopic age dating techniques, the age of the tephra and, therefore, the time of the eruption can be determined. However, in tropical environments that experience high rainfall, like Costa Rica, tephra is rarely preserved in the geologic record as it is highly erodible. Although tephrochronology can provide clues of ancient volcanism, the technique has flaws. Next, without eyewitness accounts or instrument monitoring, the magnitude of an eruption is speculative. This is especially challenging when attempting to quantify the magnitude of nonexplosive eruptions of VEI 0 to 2 because they produce little tephra. The problem becomes clear when highly effusive flood basalts need to be ranked by magnitude (Figures 8.7 and 8.8). Lastly, few volcanoes maintain the same behavior over geologic time, as these systems are not static. For example, a volcano in Iceland called Hekla (1,473 m; 63.9925, −19.6655) erupted almost like clockwork in 1970, 1980, 1991, and 2000, with an RI of approximately 10 years. Hekla has not erupted since 2000, making the volcano's RI longer than it has had for the previous four decades.

How long do volcanic eruptions last?

By definition, a volcanic eruption begins when volcanic products are brought to the surface and ends when the system's pressure and gases dissipate. This is an important distinction because the cumulation of low-level seismic energy can occur months or years prior to a volcano's paroxysmal phase. As applied by the VEI (Figure 8.1), the length of a volcanic eruption is quantified in terms of lasting hours, days, or longer; those

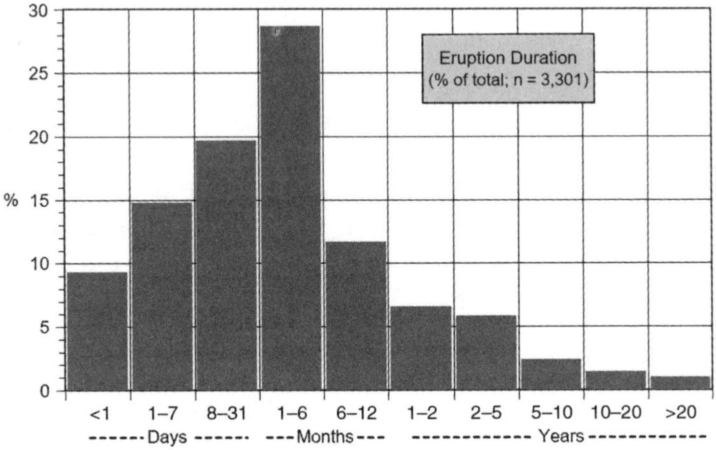

Figure 9.2 Eruption duration for 3,301 eruptions for which stop (and start) dates were documented. Modified from Simkin and Siebert (2000).

designated as cataclysmic or larger (VEI 5+) extend 6 hours or more. According to Simkin and Siebert (2000), 9% of eruptions end in less than 1 day, 14% end in 1 week, 19% end in 2 weeks, and 28% end within 1 to 6 months (Figure 9.2). The authors found that the median duration of volcanic eruptions is about 7 weeks.

Although it becomes increasingly less likely, some volcanoes have been continuously active for remarkably long durations, including Stromboli, Italy (779 m; 38.7938, 15.2111), Mount Yasur, Vanuatu (278 m; −19.5286, 169.4969), and Arenal, Costa Rica (1,589 m; 10.4633, −84.7030). Stromboli, a volcanic island in the Tyrrhenian Sea, has been erupting almost continuously for the past 2,400 years. The eruption style of Stromboli is described as moderately explosive eruptions of basaltic magma with moderate gas content. This volcano produces intermittent, explosive bursts of tephra that can be forced hundreds of feet into the air, although a sustained eruptive column does not form. This low-intensity eruption style is recognized as "Strombolian" on the VEI (Figure 8.1). Next, a volcano in the Vanuatu volcanic arc, Mount Yasur, erupted for 750 years and was sighted by Captain Hook in 1774. The largest eruptions described at Yasur are VEI 3, and it continually emits gas and ash into the atmosphere. Lastly, Arenal, a volcano that had been dormant for 400 years, erupted on June 29, 1968, and remained active for the following 42 years (Figure 9.3). The

Figure 9.3 Photo of the author, Dr. Tamie Jovanelly, and her husband, Joe Cook, in front of Volcano Arenal, Costa Rica.

initial eruption sent a pyroclastic flow down its western flank to bury two never-rebuilt villages. A town on the eastern flank, La Fortuna, was constructed to accommodate scientists and tourists who would flood the area to view the eruption and the textbook conical-shaped stratovolcano.

To date, the following 13 volcanoes have been erupting more or less continuously for the last 48 years and are projected to remain active into the future: Etna, Italy; Erta Ale, Ethiopia; Manam, Langila, and Bagana, Papua New Guinea; Semeru and Dukono, Indonesia; Suwanose-jima and Sakura-jima, Japan; Santa Maria and Pacaya, Guatemala; Sangay, Ecuador; and Erebus, Antarctica.

Can volcanoes go dormant or extinct?

Volcanoes can be active, dormant, or extinct. An active, or historically active, volcano is one that has experienced volcanism within the Holocene epoch, or the last 10,000 years. A dormant volcano can go through long periods of quiescence, with their inactivity lasting tens of thousands of years, followed by comparatively short intervals of activity. The period between eruptions is also referred to as the repose of the volcano. Most eruptions with a low explosivity (VEI 0–2) follow a repose interval of only 1–10 years. However, those volcanoes with longer repose intervals between eruptions are much more likely to be explosive (VEI 5+), which explains why most violent eruptions in the last two centuries were dormant volcanoes (Figure 3.5). Volcanoes become particularly dangerous when repose intervals exceed historical records, as unforeseen eruptions contribute to large death tolls (Simkin and Siebert, 2000). Thus, of the ~1,500 active volcanoes on the planet, most are in some phase of dormancy. Mount Kilimanjaro, Tanzania (5,820 m; 3.0663, 37.3552), and Mount Fuji, Japan (3,749 m; 35.3611, 138.7272), are two examples of volcanoes that are currently experiencing long periods of dormancy but are expected to erupt in the future.

To be clear, the designation of the Holocene cut-off between active and dormancy is an arbitrary but practical way for geologists to categorize volcanism. Similarly, an extinct volcano is one that has not erupted in the last 10,000 years and is not likely to erupt. A volcano can go extinct if it has moved away from a magma hot spot or if there has been a shift in plate tectonics. The Hawaiian-Emperor seamount chain represents several islands that have moved away from the Hawaiian hot spot over the past 70 MY and are now extinct (Figure 1.8).

What are the instruments used to monitor volcanoes?

Precursory volcanic monitoring serves two main purposes. The first reason for monitoring a volcano at a time of rest is to collect scientific data that aids in understanding Earth system processes, and the second is to apply this information to volcanic risk assessment, forecasting, and mitigation. No doubt, with each decade, new advancements in technology increase the quality, quantity, and accessibility of a variety of data sets obtainable for analysis, thereby improving scientific understanding and warning systems.

As a collective, instruments used to monitor volcanoes identify physical and chemical processes via field or, more recently, satellite observation. Physical observations, like the measurements of seismicity, ground deformation, surface temperature, and gravity anomaly, are among the original surveillance tools used by geophysicists to understand volcanology. With 100 years of seismograph deployment, the measurement of seismicity, or ground movement, is among the most commonly used predictors of an eruption within a range of short to midterm forecasting of days to weeks prior. This is because as a magmatic intrusion upwells toward the surface, the viscous fluid will infill preexisting conduits or create new ones, causing energy to be released as the bedrock fractures. This release of energy is reported as the earthquake magnitude. Before an eruption, it is not uncommon for tens of thousands of low-level ($M \leq 3$) earthquakes to be recorded as an earthquake swarm. Concurrently, as the magma continues to intrude, the ground surface will inflate or uplift. Traditionally, tiltmeters have been installed around volcanic areas to monitor elevation changes, but huge improvements in resolution have been made using synthetic aperture radar (SAR) interferometry. Using a satellite, SAR uses radar images of large areas (such as the whole volcano) and compares images taken to older images. The difference in ground surface elevation reflects inflation or the location of magma approaching the surface. Thus, identifying ground deformation can help determine where an eruption is imminent on the volcano or within the volcanic field. For example, a bulge, which ultimately would be the site of the explosion, was recorded on the northern flank of Mount St. Helens before the May 18, 1980, eruption. As one would suspect, ground deformation is usually confirmed by increases in surface temperature and vice versa. The principle of gravity monitoring begins with the idea that gravity exerts force on the surface of the Earth. Instruments called gravimeters measure subtle changes, as small as 0.00000001 m^2/s in the Earth's gravitational field around a volcano. Prior to an eruption, the gravity field around a volcano will increase as magma accumulates beneath the subsurface and decrease when magma drains away. Again, while ground deformation and surface temperature typically correlate to increases in gravity measurement, volcanic environments are sometimes surprising. Gravity measurements collected at Kilauea for the past 50 years indicate increases without significant ground surface elevation and subsidence. Geophysicists suspect that this is an indicator of void space below the volcano's subsurface.

Chemical observations of volcanic activity provide a distinct data set compared to the physical setting just described. The measurement of volcanic gases that escape to the surface via fumaroles or hot springs gives insight into the source of the magma. It helps explain how the magma approaches the surface through hydrothermal systems. As a predictive tool, volcanic gas analysis can determine whether an eruption is more or less likely to be violent, as the greater the volatile content, the more explosive an eruption will be. However, surveying volcanic gases has proven challenging for a variety of reasons. First, access to high-temperature fumarolic areas in an active crater is often difficult and hazardous. Next, the installation of equipment has proven to be semipermanent, as acidic environments quickly destroy (expensive) instrumentation, limiting the availability of continual monitoring. Lastly, gases, such as sulfur dioxide and carbon dioxide have a low solubility in magma and are often lost in degassing processes before being captured for analysis. In recent times, advancements in flight technology, like drones, have helped to solve some problems with directly measuring volatile gas collection overactive craters. Likewise, improvements in heat-resistant materials incorporated into hazard materials suits allow volcanologists to install conductors within active zones and sample soil from volcanic flanks.

The most accurate hazard assessment will include analyzing all available data sets that describe the volcano's previous or current activity. Sometimes, this can be challenging because the acquisition of physical data sets can be nearly immediate, whereas lengthy laboratory analysis may need to be completed for a conclusive understanding of a volcano's chemical signature.

Can volcanic events be predicted, prevented, or mitigated?

Using monitoring instrumentation, geologists can determine a change in a volcano's physical or chemical behavior, which can be interpreted as a warning for pre-eruption. Using this data and referring to the volcano's tephrochronology record (where available) to understand the size and scope of past events, geologists can make long-range, intermediate-range, and short-range predictive forecasts. Long-term forecasting recognizes that the volcano is active and has the potential for an eruption. Risk assessments establish recurrence intervals and probability and proactively design volcanic hazard zone maps to minimize risk before impacts of hazards are experienced (Figure 8.10). Here, risk is a measure of the financial and human costs associated with the occurrence of such events. A unitless risk reduction equation is often applied to volcanic regions to rank volcanic hazards:

Eq. 3

$$\text{Hazard} \times \text{Vulnerability} = \text{Risk}$$

Here, hazard refers to the probability of a specific volcanic hazard occurring in an area, such as a pyroclastic flow or lahar (Figure 5.1). Vulnerability is described as the degree of anticipated loss if exposed to a hazard on a scale of 0.0–1.0. Vulnerability can be assigned to innumerable variables, including humans, agriculture, buildings, economics, and so forth. The outcome, risk, is a methodical approach to determining volcanic hazard zones.

The challenge for geologists to make intermediate-range forecasts, those warnings that are issued when a volcano first becomes restless, is that an eruption can occur within days to months or not at all, making them much harder to predict than other natural disasters that have immediacy (Figure 9.4). When a volcano becomes increasingly active, surveillance will increase, emergency management plans will be reviewed, and the public will be educated about volcanic hazards.

When a volcano gives a strong indication of eruption, urgent, sort-range warnings are issued. Such warnings will be as specific as possible and may include the probable eruption site, the hazard zones in immediate danger, an estimated date, and the anticipated style of eruption. These short-range forecasts are often used to initiate evacuation, 24-hour watches, and other reduction of harm measures. Moreover, for large eruptions (VEI 5+), it may be necessary to issue global

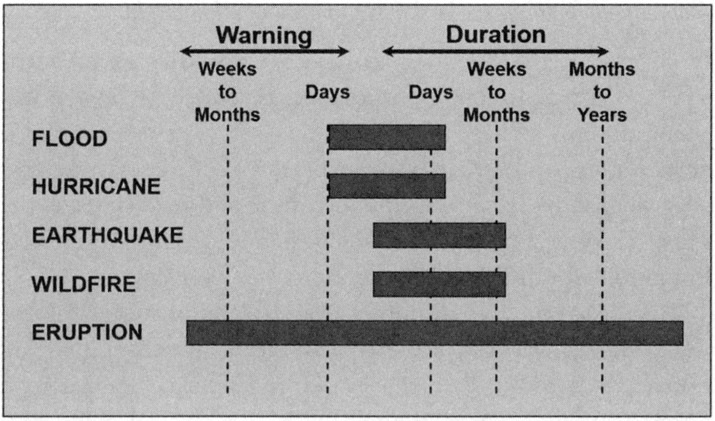

Figure 9.4 Comparison of several natural disasters to the eruption of a volcano in terms of warning time and duration of event. Modified from the United States Geological Survey, Volcanic Hazards Program.

warnings for ash or tsunamis. Once a volcano enters an eruptive phase, the activity can last from days to years (Figure 9.4). Thus, post-eruption warnings of lahars, jökulhlaups, and so forth are equally important to public safety.

While volcanic eruptions cannot be prevented, early warning monitoring has proven to save lives. A textbook example of risk mitigation is the quick response of the Philippine Institute of Volcanology and Seismology (PHIVOLCS) and the US Geological Survey (USGS) prior to the cataclysmic eruption (VEI 6.0) of Mount Pinatubo on July 15, 1991. Ten weeks before the eruption, PHIVOLCS recorded steam explosions and invited the collaboration of the USGS to install monitoring stations for seismic, ground surface elevation, and gases. On June 6, the monitoring stations recorded earthquake swarms, ground inflation, and ramping SO_2 levels. An evacuation of the area was launched on June 10 and completed before the first eruption took place on June 12. The quick efforts of the geologists saved the lives of approximately 5,000 people and protected about $250 million in military aircraft that were moved away from the eruption site.

Can humans manipulate lava flows to protect infrastructure?

In several scenarios, humans have successfully saved infrastructure by redirecting the overall pathway of lava flows. Manipulating lava is not easy, however, because magma is at its hottest on the Earth's surface at 1150°C, and it can take several months to cool completely into solid rock (Figure 3.1). Plus, the overall rate of effusiveness and volume introduces limitations for artificial interventions to divert flow.

Here listed are two examples from Iceland of successful lava diversion. On January 23, 1973, a volcano named Eldfell (88 m; 63.4347, −20.2477) on the island of Heimaey unexpectedly erupted, immediately destroying 400 homes. This volcanic island, which is 12 km off the southern coast of Iceland, remained active with an effusive lava flow for the next 6 months. The lava flows were directed at the city center and, if left uncontrolled, would infill the harbor that was essential to the local fishing industry. With help from the government, the community was able to create a complex pipeline that would allow them to spray cold seawater onto the fluid lava, causing it to cool prematurely. Thus, they were able to reroute the lava away from populated areas and kept the harbor open, saving the island's fishing industry. The eruption increased the area of Heimaey from 11.2 km^2 to 13.44 km^2.

More recently, since January 14, 2024, the Reykjanes Peninsula has experienced fissure eruptions near a city called Grindavík. This city is the largest on the southern coast, with a population of 3,500. The city supports the Svartsengi hydrothermal plant, which transports hot water to ~30,000 people living in the capital city of Reykjavik and hosts the popular tourist destination called the Blue Lagoon Spa. With low-lying topography surrounding the coastal community, it became challenging for geologists to know exactly where the lava would flow. So, in order to protect the city and infrastructure, 10-m-tall barricades were built from tephra to divert the lava flows toward the ocean. While the fabricated structures were not foolproof, they did prove successful overall.

Is it possible to safely live in an environment with active volcanoes?

Since the beginning of human settlement, volcanic regions have always been a draw for homesteading as fertile volcanic soil has high agricultural value. Today, around 8% of the world's population lives within 100 km of a volcano that has had a notable eruption during the Holocene. Thus, living in the shadows of this natural disaster risk is commonplace for almost 1 billion people.

Disaster preparedness is an important component for safely living near an active volcano. Many volcanoes in populated areas are instrumentally monitored and are evaluated for risk. With this data, geologists can make volcano hazard zone maps (Figure 8.10) that are dispersed to local communities to raise awareness of planned shelters, evacuation routes, and areas to avoid during an eruption. During an eruption, warning systems on loudspeakers, phones, and other devices relay detailed information and directions to follow. In case of evacuation, people living within an 80 km radius of a volcano are encouraged to have a Go-Kit that contains at least three days of supplies that include food, water, and medicine. If told to shelter in place, the procedure is to close and lock all doors and windows, turn off ventilation systems, close fireplace dampers, use an indoor air cleaner, and protect individuals from ash by wearing masks. Thus, the best way to address disaster is by proactively identifying potential hazards and devising safeguards to mitigate the impact.

Are the number of volcanic eruptions increasing on the planet?

Many geologists have asked, "Has global volcanism changed through time?" Although the data shows that the number of volcanoes active per year over the last 500 years has increased, this premise correlates to the growth of global population, advancements in communication, and the implementation of monitoring systems. These distinct societal changes could indicate that the reporting of volcanic events has increased instead of volcanism (Simkin and Siebert, 2000). The consistency of large eruptions (VEI 6+) being globally documented over the past two centuries helps to provide evidence for this claim.

Deligne and Sigurdsson (2015) approached this question from an extended view, using the geologic record to ask whether global volcanism is constant or episodic. These authors considered volcanism on a large scale using plate tectonic regions to define production rates at oceanic ridges, volcanic arcs, subduction zones, and hot spots. In doing so, the authors concluded that volcanism fluctuates episodically over geologic time corresponding to major plate tectonic events. Consequently, events like the assembly or break-up of Pangea (supercontinent) cycles profoundly impact magma production rates.

SUPERVOLCANO

10

CATACLYSMIC EVENTS AND SUPERVOLCANOES

How is a supervolcano defined, and how do they form?

Although originally not a scientific term, "supervolcano" has gained in popularity with recent documentaries and movies illuminating the subject as so much that it is now used by the scientific community. However, it is the preference of many geologists to refer to these events as "super eruptions" instead of "supervolcanoes." This is because volcanoes deemed "super" can have both colossal and docile eruptive phases. Silva and Self (2022) argue that "supervolcano" is a useful term if applied appropriately, as it describes the cause of an enormous eruption.

In order to be classified as a supervolcano, a volcano must meet the VEI criteria for an 8+ magnitude event (Figure 8.1) and be the site of at least one silicic explosive eruption, which eliminates continental flood basalts and large igneous provinces. This means that the ultra-Plinian eruption style will produce more than 1,000 km^3 of ejecta covering areas thousands of kilometers away from the vent with volcanic ash. The colossal explosion produces column heights that enter into the stratosphere, more than 20 km above the land surface (Figure 8.6). Additionally, the blast will be continuous for more than 10 hours. Thus, it is the enormous explosive energy that limits their recurrence interval, making them rare in the geological record (see Chapter 9, Eqs. 1 and 2).

Eruptions of these size are infrequent simply because it is unusual for magma chambers to develop, and hold, such large quantities of magma for extended periods of time. The vast chambers of molten magma that feed supervolcanoes can either form above hot spots (e.g., Yellowstone Caldera) or at subduction zones (e.g., Toba) or result from the combination of both processes. The immense volume of magma usually comes from two sources to form two distinct magma chambers referred to as the lower or deep chamber, and the upper or shallow chamber. The silica-poor (mafic) (Figure 2.3) lower chamber is located 16 to 48 km beneath the surface and is filled with magma generated from the partial melting of the subducting plate at high temperature and high depth (Figure 1.3). The melting of the sinking oceanic plate produces high-temperature basaltic magma (Figure 3.1) that pools at the base of the overlying continental crust. Because continental crust tends to be felsic, it has a lower melting temperature and begins to liquify from the heat radiated by the lower chamber to form a separate upper chamber (Figures 2.4 and 10.1 A). As the volume of magma increases in the upper chamber, the ground surface above it will inflate (Figure 10.1 B). Because the silica-rich magma held within the upper chamber has a high viscosity, and a high resistance to flow, it will trap the majority of the water and gases that are attempting to move toward the surface; however, some may be expounded at the surface as vents or fumaroles. Over time, as large volumes of magma accumulate in both chambers, the pressure of the system will exceed the magma's carrying capacity, leading to an explosive eruption. The surface above the upper chamber will collapse as explosive vents form a ring around its perimeter, causing the inflated cap to be unsupported and subside, forming a caldera (Figures 3.3 and 10.1 C). The immediate explosion of magma forces vast quantities of ash into the stratosphere and initiates rapidly moving pyroclastic flows away from the volcanic vent. Meanwhile, the newly fractured pieces of rock that fall down into the emptied upper chamber force additional magma up the outside edges of the ring. Following the supervolcano eruption, a massive caldera depression that can be more than 50 km in diameter will remain as geological evidence of the explosion.

The area within the newly formed caldera can still remain volcanically active post-eruption. Termed a resurgent caldera, this type of system is subject to eruptions of varying intensity and magnitude. Often, heat from the lower magma chamber will melt the collapsed roof, and any material that fell within the upper chamber, thereby creating a secondary magma slurry that often forms a dome in the center of the caldera (Figure 10.1 D).

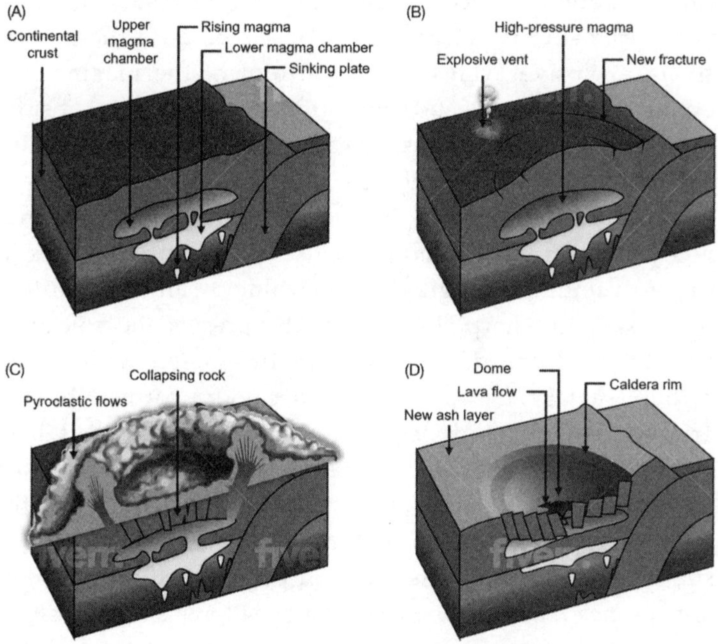

Figure 10.1 The four stages of supervolcano formation. A. Formation of upper and lower magma chamber. B. Inflation of ground surface as magma accumulates in upper chamber. C. Explosion and roof collapse as upper chamber empties forming caldera. D. Formation of dome within caldera post-eruption. Modified from Bindeman (2006).

Additionally, effusive magma from the lower chamber may rise to the surface through old vents, creating basaltic flows. Over time, however, it is possible for the system to reestablish itself, thereby infilling the lower and upper chambers with magma to generate future apocalyptic eruptions. This can result in the formation of nested, or overlapping, calderas within an area (e.g., Island Park and Henry's Fork calderas formed by the Yellowstone Hot Spot).

What does a supervolcano look like?

Unlike the steep sloping sides and prominent peak characteristic of stratovolcanoes, or the broad base of a shield volcano (Figure 3.2), the surface expression of a supervolcano is represented by a depression. This is because these massively explosive events cause the roof of the magma reservoir to collapse, and the ground above to subside, forming wide depressions, or calderas (Figure 3.3). Because supervolcanoes form calderas (>50 km in diameter) it is sometimes impossible to visually tell that you are within a volcanic basin unless you pay close attention to changes in the surrounding topography. The caldera can be identified by looking for sharp cliff faces that represent the volcano's rim. Often, these depressions can be infilled with water, forming a lake. For instance, the caldera formed by the eruptions of Toba, Indonesia, and Taupo, New Zealand, host some of the deepest caldera (crater) lakes in the world—505 m deep and 163 m, respectively. Furthermore, it can be difficult to identify caldera features because they have been subject to weathering and erosion processes that dull their appearance. The three supervolcano calderas at Yellowstone (Figure 10.2), for example, have been scoured by large alpine glaciation that occurred post-eruption. Nonetheless, nearest the caldera rim thick ash (tuff) deposits can still be admired at most supervolcanoes.

Name of Eruption and Geologic Formation	Name of Caldera	Date (Mya)
Huckleberry Ridge /Huckleberry Tuff	Island Park	2.1
Mesa Falls/Mesa Falls Tuff	Henry's Fork	1.3
Lava Creek/Lava Creek Tuff	Yellowstone	0.64

Figure 10.2 Eruption name, associated geologic formation, and date for the three supervolcano eruptions and their associated caldera name that formed from the Yellowstone Hot Spot.

How are past supervolcanoes investigated?

Discovering ancient supervolcano eruptions is a lot like being a forensic scientist, conducting an investigation through a compilation of clues left behind in the landscape. While the smoking gun for a supervolcano is finding the enormous crater, this can be initially challenging, as heavily forested or glacial landscapes may disguise the crater rims or cause them to be heavily eroded, as would exposure to water or wind. Thus, the second determining factor often found to suggest a supervolcano eruption includes analysis of volcanic ash to determine its age, its mineral composition, and its distribution. This is done by the application of geochronology, geochemistry, and petrology. Geochronology uses the decay of radioactive isotopes in magmatic minerals (i.e., crystals that solidify within molten magma, called zircons) to date volcanic processes. The age of the zircons can provide clues as to how long the magma has been stored in the chamber. Geochronology works hand in hand with geochemistry and petrology, which uses the chemistry and textures of both liquid magma and magmatic minerals to understand magma history that can describe storage depth, temperature, water content, interaction with other magma, and speed of ascent. These data can help determine the evolution of supervolcano magma chambers, along with changes in volume, composition, and heterogeneity. Moreover, understanding magma storage before and after a super eruption can provide information about resurgence. Knowledge of the size, shape, and position of magma chambers is gained through the study of geophysics. Monitoring seismic waves that are produced by earthquakes or magma movement can generate 3D images of the crust below a caldera, which can illustrate subsurface areas that are molten, partially molten, or empty. These are called low-velocity zones and can extend tens of kilometers deep.

How many supervolcanoes are there in the world today, and are they monitored?

Based on the proposed definition, Silva and Self (2022) identify nine supervolcanoes in the world today that have been active over the last 26,000 years ago (Figure 10.3). Of these, one-third are located within the United States. From the exploration of geological evidence, researchers have determined that the largest eruption ever to occur on our planet (VEI 9.1) took place 74,000 years ago at Toba, Indonesia (1,593 m; 2.627, 98.8197), and this supervolcano is still active today (Figure 8.9). The authors recognize that this list may not be complete and that there may be supervolcanoes that remain unidentified, especially in Eastern Russia and South Africa. Interestingly, Silva and Self (2022) do not include the Campi Flegrei, Italy (96 m; 40.8275, 14.1389) caldera complex on their list, leaving Europe without a supervolcano on its continent. While a significant eruption did occur at Campi Flegrei 37,000 years ago, leaving behind a 15 km caldera, it did not produce nearly the volume of ejecta (2.5 km^3) required of a VEI 8+ eruption. Nonetheless, the Campi Flegrei caldera complex, which hosts the infamous Mount Vesuvius, is still considered active and emits high levels of sulfuric gases.

Like other active volcanoes, all of the nine supervolcanoes are monitored for activity in some capacity. Volcano observatories act as clearinghouses for the collection of real-time physical data for earthquakes, ground deformation, stream flow, and temperature. Additionally, chemical signatures, such as gas flux, and the geochemistry of groundwater and surface water conditions are also often observed.

Supervolcano	Location	Last *M* 8	DRE volume (km^3)	M	Last known eruption reference	DRE volume (km3)
Toba	Sumatra Indonesia	74 ka: Youngest Toba Tuff	~ 5300	9.1	54.5 ± 8 ka; Lava flow at Pusik Buhit	< 1
Yellowstone	Wyoming, USA	631 ka; Lava Creek Tuff	~ 850	8.3	75 ka; Pitchstone Plateau Flow	70
WhakamaruTaupo	North Island, New Zealand	25.5 ka; Oruanui eruption	~ 530	8.1	1800 yrs B.P.; Taupo eruption	30
Aira	Kyushu, Japan	~ 30 ka; Osumi-Tsuyama eruption	> 430	8.0	2021 CE; Sakurajima volcano	<<<0.1
Kikai	S. Kyushu, Japan	7.3 ka Akahoya eruption	> 500	8.1	Active lava dome	~32
Aso	Kyushu, Japan	87 ka; Aso-4 eruption	< 592	8.2	2021 CE; Nakadake	<<<0.1
Long Valley	California, USA	760 ka (Bishop Tuff Hildreth and Wilson, 2007)	~600-650	8.2	16 to 17 kyr; western mafic centers	<1
Valles	New Mexico, USA	1.2 Ma; Otowi eruption	<550	8.1	74–65 ka; Banco Bonito Rhyolite	<10
Atitlán	Guatemala	74 ka Los Chocoyos	~510	8.1	67 +11/−9 ka, I-Fall	~7)

Figure 10.3 Table of known active supervolcanoes. Modified from Silva and Self (2022).

How many super eruptions have occurred, and what is their recurrence interval? How often do super eruptions occur?

Considering that the VEI scale is logarithmic, an explosion ranking 8 is considered to be an apocalyptic event. For reference, a VEI 8 is about a thousand times larger than the 1980 eruption of Mount St. Helens, Washington, and has the equivalent force of a small asteroid colliding with the Earth's surface. Thus, these large-scale events occur infrequently. From investigating the tephrochronology record around supervolcanoes it has been determined that there have been 42 colossal eruptions in the last 36 MY. Using these data, if we take an average, it can be estimated that a super eruption transpires once every 857,142 years or so. The most recent super eruption (referred to as Oruanui) was at Taupo, New Zealand (402 m; −38.6775, 176.0697), approximately 26,500 years ago, which was documented by volcanic ash found in ice cores collected in West Antarctica. So, when recurrence interval (Chapter 9, Eq. 1) and probability (Chapter 9, Eq. 2) are considered, there is a finite chance of a super eruption happening in this lifetime. However, it is commonplace for supervolcanoes to experience smaller silica-rich or effusive lava eruptions. Using Taupo as an example, since its super eruption 26,000 years ago it has had 25 eruptions ranking VEI 3 to 6. This includes a well-documented historical eruption occurring 1,800 years ago that measured VEI 5 and was accompanied by a large-scale earthquake of magnitude 6.9. The Taupo Eruption produced a 50 km column of ash that covered the entire island of New Zealand in at least 1 cm, and some areas within the blast zone were covered with 100 m of pyroclastic flow.

What is the largest known supervolcano eruption?

To date, the largest known supervolcano eruption occurred at Toba, Indonesia (1,597 m; 2.8191, 98.6077) approximately 74,000 years ago. Although the remains of the crater were mapped in 1949 by a Dutch geologist, Rein van Bemmelen, the catastrophe was only identified in the 1990s when geologists, geneticists, and anthropologists working on different problems realized that they were identifying the same global disaster. The silica-rich eruption of Toba, which was 1000X as large as Tambora or Krakatoa (Figure 8.4), is considered to be the worst explosion ever to have occurred on the planet, ranking at VEI 9.1 (Figure 8.1), and is the largest eruption in the last 2.1 MY. Moreover, the eruption created a crater 30 km wide and 100 km long that can be seen from space, produced an ash column exceeding the lower limit of the stratosphere, and released more than 2,800 km^3 of magma. Petrological analysis of the Young Toba Tuff, which is 600 m thick inside the caldera, suggests that the eruption lasted 9–14 days. The tephrochronology record of Toba found in ice and sediment cores indicates that more than 1% of the Earth's surface was covered in ash at least 10 cm deep (Rose and Chesner, 1990). In fact, the ash deposits from Toba are even found 7,000 km away from the eruption site in sediment cores collected from the bed of Lake Malawi (Lane et al., 2013). Although of lesser magnitude (VEI 8.0), Toba erupted two times prior to the super eruption: at 840,000 and 1.2 MYA.

The caldera created by the super eruption is now infilled with water, forming the largest crater lake in the world, Lake Toba. In the center of the caldera is a resurgent dome, referred to as the island of Samosir, which dates to the 74,000-year event.

Among the most discussed supervolcano eruptions are those created by the Yellowstone Hot Spot. What is interesting about its geologic history, and how many super eruptions have taken place?

The first signatures of the Yellowstone Hot Spot (YHS) were identified around 16.5 MYA along the Snake River Plain, Idaho. The origin of the hot spot remains a topic of discussion, primarily due to its sudden appearance in the geologic record. The two leading hypotheses include hot spot formation as the result of magma upwelled by convection cells formed in the upper mantle at the lithospheric boundary, or upwelling of deep-seated magma found in the lower mantle (Figure 1.1). Nevertheless, the intense heat generated by the YHS has produced large eruptions of both basaltic and rhyolitic magmas (Figure 2.3). Since its inception, the North American plate has moved over top of the YHS in an overall southwesterly direction at a rate of 2.3 cm/yr, leaving a hot spot track that has been isotopically dated (Figure 10.4).

As Figure 10.4 depicts, the YHS track leaves a trail of eruptions over a distance of ~750 km, forming nine calderas across Idaho and into other neighboring states. That is, instead of textbook peaks of stratovolcanoes greeting visitors, like Arenal in Costa Rica (Figures 3.2 and 9.3), these significantly explosive events cause the roof of the magma reservoir to collapse, and the ground above to subside, forming wide depressions, or calderas (Figure 3.3). The YHS is currently 800 km in diameter and is located at the northeastern corner of Idaho, at the Montana/Wyoming border. Geologists have

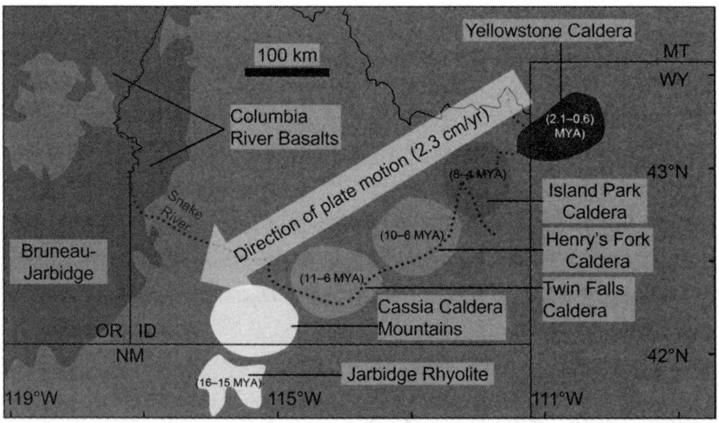

Figure 10.4 Yellowstone Hotspot Track and correlating geological history of caldera formation. Modified from Branney et al. (2008).

identified two magma chambers within the YHS; the deeper chamber (19 to 48 km beneath the surface) is about 4.4 times larger than the shallower chamber (5 to 16 km below the surface) having a volume of 46,000 km^3 (Huang et al., 2015). The enormity of the lower chamber helps to explain the more frequent basaltic eruptions that have transpired since the last super eruption 640,000 years ago.

While all of the calderas are associated with large eruptions (VEI 4+), it has only been over the past 2.1 MY that the YHS has generated super eruptions. Namely, there have been three super eruptions, which are described by their eruption name, associated geologic formation, or caldera name (Figure 10.2). Tuff, a combination of hot ash, pumice, and other rock fragments welded together, is the most commonly associated geologic formation linked with Yellowstone volcanism.

The oldest super eruption occurred 2.1 MYA forming one of the world's largest calderas, Island Park, Idaho (1,901 m; 44.3311, −111.3288), which measures 80 km long by 65 km wide. For perspective, the caldera occupies enough area to include the sprawling city of Tokyo. The ashfall, referred to as the Huckleberry Tuff, covered a 2,450 km^3 area and can be found from southern California to the Mississippi River near St. Louis (Figure 10.5). In close proximity to the caldera, the Huckleberry Tuff can be as thick as 400 m. The only larger eruption currently known is that which occurred at Toba, Indonesia, approximately 74,000 years ago, producing 2,800 km^3 of ash (Figure 8.4).

The next cataclysmic eruption of the Yellowstone Hot Spot occurred 1.3 MYA, forming the Henry's Fork Caldera (29 km long by 37 km wide). Smaller, the Henry's Fork Caldera is nested within the Island Park Caldera. The eruption produced 280 km^3 of ash known as the Mesa Falls Tuff, covering a region including all of Wyoming, most of Colorado, and parts of Nebraska and Kansas.

At approximately 640,000 years ago the last supervolcano eruption happened to form the Yellowstone caldera, Wyoming

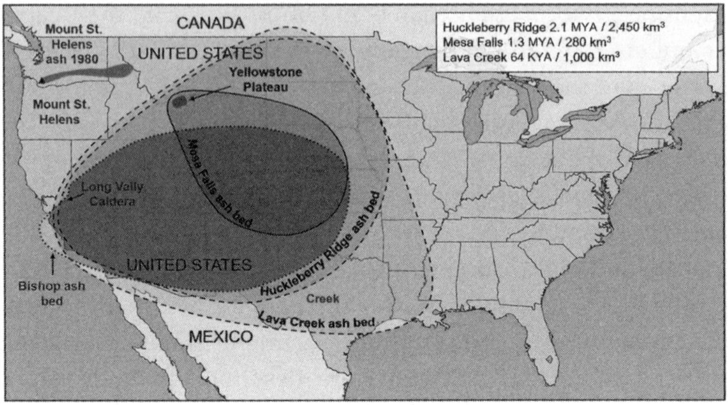

Figure 10.5 Map of supervolcano eruptions in the United States and their associated ashfall deposition.

(2,596 m; 44.4125, −110.7230). While smaller than Island Park, the Yellowstone caldera measures 70 km long by 45 km wide and produces 1,000 km^3 of ash forming the Lava Creek Tuff, which blankets half of the United States (Figure 10.5).

An eruption that took place 760,000 years ago formed the Long Valley caldera (16 km long by 32 km wide) located in eastern California (2,342 m; 37.7116, −118.8841). Considered a supervolcano by some, this eruption produced 600 km^3 of rhyolitic ash that formed the Bishop Tuff, which is a significant deposit found in southern California (Figure 10.5). This eruption, however, is not linked to the YHS. Instead, it is linked to magmatic upwelling occurring along the Pacific Plate subduction zone (Figure 1.5).

Is the Yellowstone Hot Spot still volcanically active, and will it produce another supervolcano eruption?

The area known as the Yellowstone Plateau Volcanic Field (YPVF) (Figure 10.5) encompasses the three supervolcano calderas (i.e., Island Park, Henry's Fork, Yellowstone) produced by the Yellowstone Hot Spot over the last 2.1 MY and remains volcanically active today. Since the last supervolcano eruption occurred 640,000 years ago to form the Yellowstone Caldera, there have been approximately 80 nonexplosive, basaltic, eruptions. The most recent of these was a 70 km^3 effusive flow on the Pitchstone Plateau, Wyoming (44.2741, −110.7986; 2,720 m), that happened 70,000 years ago. Additionally, since the last ice age ended approximately 15,000 years ago, the YPVF has become increasingly hydrothermally active due to the isostatic uplift of the ground surface, producing more than 25 hydrothermal explosion craters of 100 m in diameter or more. A hydrothermal explosion occurs when a subsurface cavity is infilled with water and comes into contact with a heat source (magma) that brings the water up to its boiling point whereby creating steam. Eventually, the pressure in the cavity becomes too great and an explosion will occur, leaving a crater on the surface. Once the explosion materializes, and the cavity is empty, the crater becomes dormant. With the recurrence interval of hydrothermal explosions being every 200–700 years, this risk is deemed the most significant hazard in the YPVF. This frequency is much shorter than the 700,000-year recurrence interval established for another supervolcano eruption at the Yellowstone Hot Spot (YHS), which would equate to a 0.0000001% probability of an eruption of equal size occurring each year. This is further confirmed by the shallow, rhyolitic, magma chamber of the YHS only containing 5%–15% molten material, which would not be enough magma to supply a super eruption. Additionally, the Yellowstone Caldera continues to subside at a rate of 2.5 cm per year, indicating an empty shallow magma chamber.

It is anticipated that the YHS will continue to provide heat to support the geothermal activity in the YPVF well into the future. This is critical to Yellowstone National Park's most visited feature, the Old Faithful Geyser (2,242 m; 44.4605, −110.8280). Every 90 minutes a steam eruption occurs, bringing 3,700–8,400 gallons of water to the surface as a column that reaches ~90 m into the air lasting between 1 and 5 minutes. Likewise, the park currently includes some 10,000 other thermal features including mud pots, fumaroles, and hot springs.

Over geologic time, as the North American Plate continues to move along a southwesterly course, the YPVF will become dormant, and the hot spot will create a new volcanic field. Craters and geothermal activity in the YPVF will provide geologic evidence of past hot spot activity, as does the 640 km rock record of Snake River Plain, Idaho (Figure 10.4).

What are the effects on the landscape when a supervolcano erupts?

The direct measurement of a supervolcano does not exist because there has not been this level of volcanic activity on the planet since Toba, Indonesia, erupted (Figure 8.4) 74,000 years ago. However, instrumentally monitored modern examples of Hunga Tonga–Hunga Ha'apai and Mount Pinatubo, Philippines, give examples of how the planet can change when large-scale (VEI 6.0) volcanism occurs. During these eruptions, which were 1000X smaller than a supervolcano, the globe experienced short-term (months to years) temperature changes largely resulting from the influence of water vapor and sulfuric acid.

While firsthand observations are impossible for past supervolcano events, the application of a variety of different climate proxies allows geologists to speculate what changes to the environment, namely land and ocean temperatures, result from these apocalyptic eruptions. Often, the analysis of ice and sediment cores is used to describe super eruptions, as these cores are known to preserve volcanic ash and isotopic data very well, and they help to provide a global signature.

For example, evidence of the last super eruption of Toba is found in Greenland sediment cores. Specifically, extreme levels of sulfate aerosol deposits which had settled out of the atmosphere were identified in cores at 74,000 years ago (Zielinski et al., 1996). Petrological analysis of the 74,000-year-old solidified rock at Toba confirmed how much gas was released during the eruption, agreeing with extreme sulfate emittance (Rose and Chesner, 1990). The chemical, sulfate, goes through a reaction in the atmosphere to produce sulfuric acid (H_2SO_4), which blocks incoming solar radiation, or heat, from reaching the surface of the Earth. Using the sulfate emission estimates provided by Zielinski et al. (1996), other researchers were able to calculate a decrease in global temperature between 3.2°C and 4.5°C at the time of the eruption (Oppenheimer, 2002). Similarly, oxygen isotope ratios extracted from marine fossils found in sediment cores were used to reconstruct past sea surface temperatures. When this proxy analysis was completed for sediment cores collected in the China Sea, researchers found that after the Toba super eruption, sea surface temperatures became colder by 1°C (Huang et al., 2001). Thus, a variety of different lines of evidence suggest that the eruption of the Toba supervolcano 74,000 years ago shifted the global climate into a severe cold phase for 6 to 10 years and contributed to a short ice age, lasting ~1,000 years, that is referred to as the Greenland stadial 20 or GS20.

Like Toba, it is thought that eruptions at other supervolcanoes result in the "global volcanic winter" phenomenon. The effects of a supervolcano eruption can be long-lasting because a reduction in temperature at the Earth's surface can cause an albedo effect. The albedo effect occurs when polar regions become colder, encouraging ice sheets to grow. When this happens, the white surface of the ice sheets reflects the incoming solar radiation from the sun (instead of absorbing the heat as a darker, ground surface would). This creates an amplifying or positive feedback cycle that promotes ice sheet growth and results in overall cooler temperatures until something in the environment counterbalances the action.

Additionally, another effect of the immense emissions of sulfur dioxide (SO_2) during a super eruption is the depletion of the ozone layer. The ozone layer forms about 10 km above the surface within the stratosphere (Figure 8.6) and absorbs heat reaching the Earth from the Sun to provide a suitably heated planet for life to thrive. The ozone layer becomes damaged during super eruptions and cannot regenerate itself because photons of ultraviolet radiation require sunlight to form, and the sunlight is inhibited by gases and ash floating in the stratosphere. With a depleted ozone layer, the climate becomes cooler because heating from the Sun escapes back into space. With Toba, researchers believe that the super eruption caused severe tropical stratospheric ozone depletion by as much as 50% (Osipov et al., 2021).

ANCIENT ERUPTIONS

11

HOW DO GEOLOGISTS INVESTIGATE MODERN AND ANCIENT ERUPTIONS?

What are the foundational principles that are applied to dating techniques in geology?

In geology, dating techniques refer to methods used to determine the age of rocks, sediment, volcanic ash, or geologic features. These techniques are not just theoretical concepts, but practical tools that help us understand the history and evolution of our planet. The first geologic time scale (Figure 1.2), formulated in the late 1800s, was established using approximate positions of rock strata in chronological order. More than a century later, technology advanced so that exact ages could be assigned to rocks, completing the geologic time scale. Today, dating techniques are commonly divided into two categories—relative and absolute—and they are often used in conjunction.

Relative dating, or relative time, is looking at an object (or a layer of ash, soil, or rock) to identify its stratigraphic position compared to those above and below without providing a specific date. The purpose is to look at the sequencing of the rocks relative to one another. It was established in the 17th century by Danish geologist Nicolas Steno, centuries before instruments were invented that could assign specific ages. Several principles of stratigraphy apply relative dating techniques, including superposition, original horizontality, lateral continuity, crosscutting relationships, inclusion, and faunal succession based on fossil assemblages. Relative dating techniques begin with the concept of superposition, which describes how younger strata (layers) are deposited on top of older ones. Thus, if looking at a rock outcrop in cross section, it can be assumed that the oldest layer is always on the bottom unless something unusual (i.e., folding or faulting) occurs to change its position (Figure 11.1 A). The law of original horizontality suggests that when strata are deposited initially, they will blanket a surface horizontally (Figure 11.1 B, left). Suppose a volcano erupts, and ash goes into the atmosphere. In that case, it will fall back to the surface to cover everything within range, so if the ash layer in the outcrop appears tilted or folded, something happened in the environment to cause such change *after* the layer was deposited (Figure 11.1 B right). The principle of lateral continuity is often applied to areas that have experienced weathering and erosion due to river systems (or glaciers), creating a channel, gorge, or canyon. This principle assumes that because the strata were initially deposited in a uniform horizontal layer, if they were missing, they were eroded (Figure 11.1 C). So, if you are standing on one side of the riverbank and looking across the channel, you can assume that the same stratigraphy lies beneath your feet. The law of crosscutting relationships helps to identify the relative age of dikes and plutons compared to other strata they intrude into (Figure 11.1 D). Here, looking at an outcrop from bottom to top (oldest to youngest), the strata must be in place before magma can be injected into a fracture. Therefore, the dike or pluton is relatively younger than the rock surrounding it. This concept is the opposite of the law of inclusion: pieces of older rock can break off with magma upwelling and become included within. Thus, the inclusion (rock fragment or xenolith) is older than the rock surrounding it (Figure 11.1 E). Lastly, the fossil record can be used to correlate strata from one location to another using an index fossil or fossil assemblage that consistently appears in the same geological time frame (Figure 11.1 F). A trilobite is an example of an index fossil; it is easily identifiable, geographically

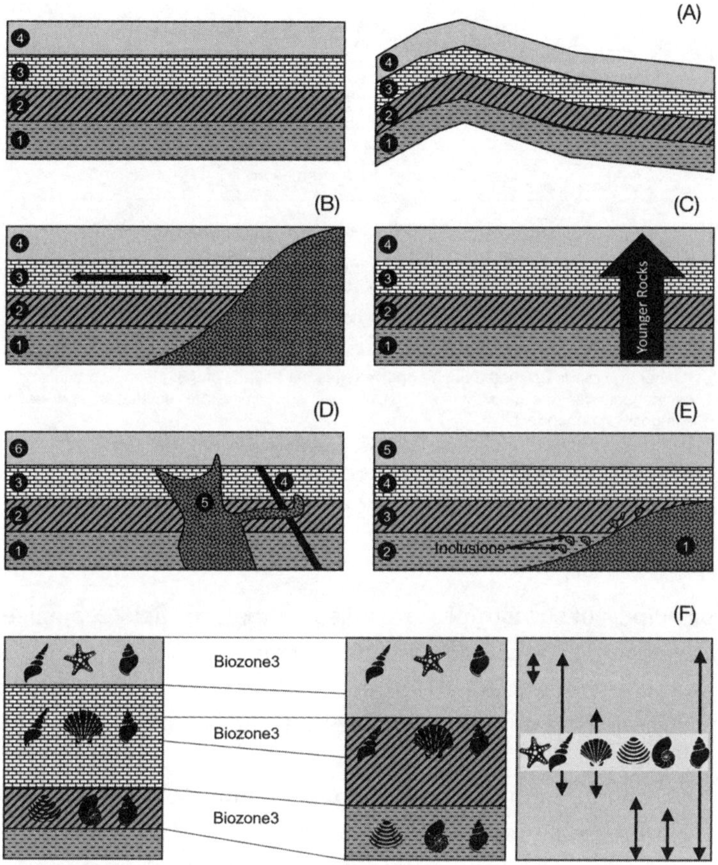

Figure 11.1 Principles of stratigraphy. A. Original horizontality. B. Lateral continuity. C. Superposition. D. Cross-cutting relationships. E. Inclusion. F. Faunal succession based on fossil assemblages.

widespread, well-preserved, and lives in a constrained time frame (Cambrian to Permian Periods). Thus, if a trilobite is found in a rock stratum, then the age of the rock must be between 540 MY and 250 MY. A similar application can be made for a group of fossils. Although volcanic rocks do not contain fossils, the principle of fossil succession can be applied above or below igneous strata. While seemingly primitive, these

Technique	Meaning of age	Age of material that can be dated*
$^{40}Ar/^{39}Ar$	Timing of eruption	5,000 to 6.25 billion years
U-Pb (Uranium series)	Timing of crystal formation	100,000 to 22.5 billion years
^{238}U-^{230}Th (Uranium series)	Timing of crystal formation	5,000 to 350,000 years
^{230}Th-^{226}Ra (Uranium series)	Timing of crystal formation	0 to 8,500 billion years
Cosmogenic (^{36}Cl, ^{10}Be, ^{21}Ne, ^{26}Al, ^{3}He)	Duration of surface exposure	0 to several million years
^{14}C (Carbon dating)	When carbon uptake stopped	0 to 50,000 years

*Upper age range estimated based on five times the half-life of the daughter product or cosmogenic isotope (except for ^{14}C). Lower age limits are estimates and are composition dependent.

Figure 11.2 Table showing different types of geochronology techniques, the ages over which those techniques are best applied, and the meaning of the ages determined by the techniques. Modified from United States Geological Survey, Public Domain.

principles of stratigraphy are essential tools for field geologists attempting to qualitatively understand landscape processes and their application aids in the creation of an initial timeline of events that later can have dates assigned by absolute dating techniques.

Absolute dating techniques provide precise dates for geological materials or fossils. At its simplest, the method can be applied to the physical counting of annual growth represented by tree rings (i.e., dendrochronology) or growth bands found in some corals, as seasonal deposition of sediment recorded as layers found on lake or ocean floor bottoms, or as records of annual precipitation identified in ice cores. While important and applicable globally, the time encapsulated by physical counting methods is limited and does not extend past the Quaternary Period. Thus, other methods become necessary to date older events or the age of the Earth absolutely.

With the technological advancement of the mass spectrometer in the 1950s, scientists could accurately measure radioactive isotopes found in organic and inorganic materials (Figure 11.2).

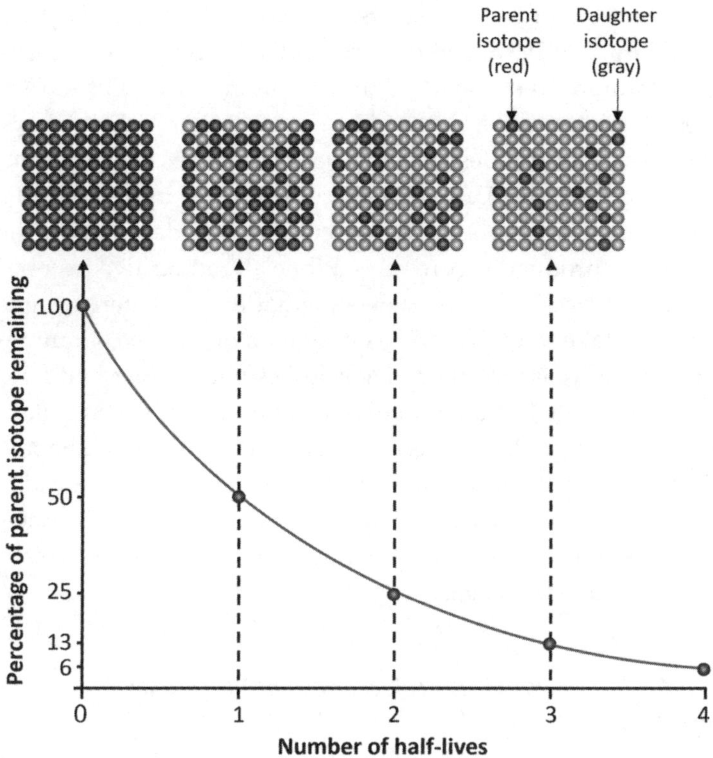

Figure 11.3 Isotopic decay chart comparing the number of half-lives to the percent of parent isotope remaining over time.

By definition, an isotope is one of at least two forms of the same chemical element that contains atoms that have the same number of protons but with a different number of neutrons. An atom has protons and neutrons in its nucleus that have mass that a mass spectrometer can weigh. While some isotopes are stable, radioactive isotopes decay over time, emitting energy (i.e., radiation). Radioactive isotopes decay from a "parent" element into a "daughter" element (Figure 11.3). As the decay process continues, the daughter product approaches stability. An example of an element that has isotopes is Uranium. On the periodic table of elements, Uranium has the atomic number 92 and the atomic weight 238.0. The parent isotope, Uranium-238, decays into its daughter isotope Uranium-235 (or Uranium-234 as part of the Uranium disequilibrium series). Uranium-235 has three less neutrons than Uranium-238, so it is lighter (has less mass). The time it takes for Uranium-238 to decay into Uranium-235 is measured by its half-life. In one half-life, 50% of the parent product decays into the daughter product (Figure 11.3). While all radioactive isotopes have half-lives that follow exponential decay patterns, some isotopes take much longer to decay than others, like Uranium (4.5 billion years). Radioactive isotopes with longer half-lives can be used to date older material and those with shorter half-lives younger. This process is described later in the chapter. The geologist's job is to choose the appropriate radioactive isotope for the age-date materials. As a check and balance, geologists often use more than one radioactive isotope to confirm and constrain ages.

What is the science of tephrochronology, and how is it used in volcanology?

First, tephra, a Greek word for ash, is formed by the fragmentation of primarily silica-rich magma during explosive volcanic eruptions, ejected into the atmosphere (or higher) in tall eruption columns, allowing for the less dense material to be dispersed widely in the prevailing wind direction. The grain size of tephra refers to the diameter of individual tephra particles, with the smallest being referred to as ash (<2 mm), followed by lapilli (2 mm to 64 mm), and the largest classification, volcanic bomb (>64 mm). Tephra blankets an area (Figure 11.1 B left), creating a "tephra layer" that can be extremely widespread, covering thousands of kilometers on land and in ocean basins. The thickest portions of the tephra layer will be nearest the eruption site and thin distally. Consequently, the thickest portions of the tephra layer usually represent larger (denser) particle sizes.

Tephrochronology, a stratigraphic method, uses tephra ash from individual volcanic eruption events to determine a chronological order to reconstruct geologic time, ancient environments, or archaeological events. The origins of tephrochronology began in Iceland, where it was first used in the 17th century to describe soil sections. Centuries later, in the 1930s, an Icelandic doctoral student, Sigurdur Thorarinsson, began to observe relationships between pollen and tephra dispersal patterns. In his dissertation, he coined the term "tephrochronology." This tool was initially applied as a relative dating technique or a method for determining the order in which events occurred in the past without assigning a specific age to each event. However, tephrochronology has developed into an important absolute technique—using isotopic dating techniques, firm dates can be determined and tied to geochemical signatures of specific volcanoes, thereby expanding its use to determine the distal extent of volcanic ash and volume. Isopach maps (e.g., Figures 8.2 and 8.9) are common representations of tephra distribution that are based on land surveying (for recent events) or investigation of soil profiles (historical events). Likewise, the methods of tephrochronology can be applied to ice and soil cores. The overall dispersal of tephra is influenced by environmental factors that include explosion type and duration, plume height, prevailing wind direction, presence of water, and strength at the time of eruption.

Although it is not an exact science, tephrochronology can aid in determining the volume of ejected material. Volumes of tephra are commonly published as bulk volume or compacted volume, freshly fallen or uncompacted volume, and dense-rock equivalent (DRE). Additionally, in areas where tephra has fallen offshore, two volumes may be listed: on land and sea. For events not recorded during historic times, estimating the volumes can provide an indicator of explosivity that can be correlated to the Volcanic Explosivity Index (VEI) (Figure 8.1).

Tephra layers are often sought out as marker beds in ice and sediment cores. A marker bed is easily distinguished by its physical characteristics, such as grain size, color, or magnetic properties. Because tephra layers form during volcanic eruptions, they instantaneously become part of the geologic record and provide reliable time constraints for layers above and below. They also provide a means to correlate stratigraphy from one place to another.

How does volcanic ash become preserved?

Volcanic ash can be preserved as a consolidated material, called tuff, or it can be altered by processes of weathering and erosion, whereby it turns into soil. Or, as is the case with ice cores, ash is simply preserved as ash. All of these preservation processes depend on the amount of ash initially deposited, the size of the ash particles, the presence of water, and the overall climatic conditions.

Volcanic tuff is formed when volcanic ash is deposited in an environment and becomes compacted through settling or burial processes over time. The pieces of ash become lithified together, turning the loose material into a solid form generally classified as sedimentary rock. During the initial volcanic episode or soon after, ash can be welded (melted) together to form a solid mass more quickly. It is not uncommon for tuff to be a remnant of a pyroclastic flow.

Volcanic ash can decay into soil as a result of physical and chemical weathering processes experienced in the natural environment with exposure to water, wind, or ice. These environmental factors will break down the ash particles, releasing essential minerals and nutrients into the subsurface to create a fertile soil type called an andisol. While steep volcanic slopes may limit their arability, valleys containing andisols are favored for farming because they are permeable, erosion-resistant, readily tilled, and have a high mineral and nutrient content.

Although volcanic soil only covers 1% of the Earth's surface, andisols are important to civilization, making agricultural practices possible that feed about 10% of the world's population (Ping, 2000). From early humans surviving in the Rift Valley of East Africa to coffee plantations established by Mayan and Inca cultures in Central and South America or historic vineyards still growing today in the Mediterranean, it is the specific properties of andisol soil that anchored families in volcanic regions.

How are volcanic deposits absolutely dated?

As described in earlier chapters, the composition of the magma will largely determine which type of volcanic material is produced at the Earth's surface. No matter whether the resulting material solidifies as a rock or volcanic ash, many accurate and precise techniques can be used in age determination. The most common dating techniques for volcanic material include isotopic (radiometric) dating, thermoluminescence, cosmogenic radionuclide surface exposure, radiocarbon dating, paleomagnetism, and lichenometry.

Isotopic, or radiometric dating, measures the rate of decay of the original isotope (called parent) to its stable byproduct element (called daughter) (Figure 11.3). The relative proportions of parent and daughter atoms, measured by a mass spectrometer, can be used to determine the age of the material in which they are contained. The decay of Potassium-40 (K-40) to Argon-40 (Ar-40) can be applied to most volcanic material. However, the initial amount of potassium can differ depending on the rock type, with some types, like rhyolite, having higher content compared to basalt (Figure 2.3). The radiometric clock begins at the time of lava solidification when the parent material (K-40) begins to decay into its daughter product, Ar-40. The ratio of K-40 to Ar-40 revealed by the mass spectrometer will determine how long ago the lava solidified, which is also the age of the rock; the more significant the amount of Ar-40, the older the specimen (Figure 11.3). Because the half-life of K-40 is relatively long (1.25 billion years), this dating technique is only useful for rocks 2.6 million years (Pleistocene-age) or older. This limitation was overcome when a more sophisticated technique was developed using Argon-40 and Argon-39. Known as argon-argon dating, this approach irradiates stable potassium-39 (K-39) found present in the volcanic sample with neutrons in a nuclear reactor to create a small amount of argon-39 (Ar-39). The proportion of the reactor-made isotope of Ar-39 is then compared to the Ar-40 naturally present in the sample to yield an age. Argon-argon dating is considered one of the most precise methods for dating volcanic rocks, allowing scientists to pinpoint the timing of eruptions with high accuracy. Additionally, this technique has expanded the age-dating capability of argon to volcanic material as young as a few thousand years old. As an example, when argon-argon dating was used to analyze the volcanic ash from the historically documented AD 79 eruption of Mount Vesuvius, Italy, the mass spectrometer revealed a date with a 1% error margin (Showstack, 2011). In some volcanic landscapes, it may be appropriate to combine isotopic dating techniques. Along the Yellowstone Hot Spot track (Figure 10.4), for example, potassium-argon, argon-argon, and uranium isotopes have been analyzed. This helped to age not only young lava flows but also xenoliths, which were dated to inform geologists how long the magma was sitting in the chamber before erupting (Figure 2.5).

Another way to assign ages to volcanic material is through the thermoluminescence (TL) dating technique. TL measures the time that has elapsed since a crystalline material, namely quartz, was last heated or exposed to sunlight. This method can be applied to volcanics because felsic and mafic magmas contain some silica (Figure 2.3). This approach measures the amount of radiation a crystal has absorbed and stored, which is then released as light when the sample is heated under laboratory conditions. TL is often used to determine the time that it took a lava flow to cool, or it can be applied to an underlying stratum that was last exposed to sunlight before being covered by a lava flow. Thus, the main applications of TL include (1) establishing the chronological order of multiple eruptions from a single volcanic source, (2) identifying the age of tephra layers to correlate other geologic events, and (3) age date surrounding paleoenvironmental changes. TL is most accurately applied to volcanic rocks formed in the late Pleistocene, or between 10,000 and 250,000 years ago, although younger and older samples may be analyzed. The early application of TL in Hawaii informed geologists of the ages of at least 30 lava flows, which had previously been only described by archaeological evidence. This led to the first conceptualization of the Hawaiian Island Seamount Chain (Figure 1.8; May 1979).

Next, the surface of the Earth is constantly exposed to cosmogenic radionuclides (CRN) that originate from high-energy supernova explosions in space. A supernova is a massive explosion that occurs when a star collapses. The CRN (i.e., beryllium-10, aluminum-26, chlorine-36, carbon-14) expelled in space are absorbed by all things on the surface of the Earth, both inorganic and organic. By measuring the concentration of CRN stored within a volcanic rock, geologists can determine how long that rock has been exposed to cosmic rays, which correlates to the time of eruption. In addition to CRN being used to reconstruct the sequence of eruptions, it can also be used to provide information on how long a surface has been buried, how quickly volcanic areas are eroded, and in volcanic areas affected by glaciers, CRN can be used to date glacial deposits and understand past glacial advances. CRN dating is applicable over short to long timescales (1,000–10 million years) with variability depending on which isotope is applied. Because CRN is absorbed only in the first meter of the Earth's surface, great care must be given to the sampling method to arrive at the most accurate dates.

Carbon-14 (C-14) meets the definition of a CRN because it is primarily created in the Earth's atmosphere through the interaction of cosmic rays with Nitrogen atoms. This process produces C-14 found in the atmosphere, which is absorbed by plants during photosynthesis and other life unambiguously by eating plants or animals. That is, all living organisms contain C-14, and upon death, the C-14 (parent isotope) will decay into Nitrogen-14 (daughter product) over a half-life of 5,730 years (Figure 11.3). This isotopic decay process is referred to as radiocarbon dating. While radiocarbon dating can be applied to anything that contains carbon that is less than 60,000 years old, it is primarily utilized in volcanic regions to indirectly date volcanic eruptions by analyzing the carbon content of charcoal that was buried by volcanic ash during the eruption. Charcoal is a highly sought-after proxy because it can form during volcanic eruptions when hot lava flowing at the surface initiates forest fires or the like.

Paleomagnetism uses magnetic characteristics to determine the age of volcanic rocks. Igneous rocks, namely those mafic like basalt (Figure 2.6 C), are helpful for paleomagnetism because they contain elevated amounts of magnetic minerals, like iron (Figure 2.3). When magma erupts to the Earth's surface in its fluid state, its temperature is between 800°C and 1200°C (Figure 3.1). As the magma solidifies to lava, magnetic minerals within will orient themselves toward the Earth's magnetic north pole, thereby "trapping" the compass direction. While the location of Earth's magnetic north changes over time with the movement of the tectonic plates, the average position of its wander always coincides with the geographic north pole. The magnetic signature of iron-rich rocks can be measured by an instrument called a magnetometer. In some locations, the paleo location of the magnetic north pole is well known as it corresponds to geological time. Otherwise, only the relative age of two rocks can be determined by measuring and comparing the magnetic directions recorded in the rocks. If the rocks are the same age, they will point to the exact magnetic north pole location. If the rocks solidify at different times, they will point to different magnetic north poles. Thus, determining a rock's paleomagnetism can estimate the age of volcanism by identifying which period in the magnetic field history it matches. This technique is often used with other dating techniques to refine ages to a finer scale.

Lichenometry is a technique used to relatively date the age of rock surfaces, deposits, or other substrates by measuring the size or abundance of lichens. By definition, a lichen is a symbiotic organism formed by a mutually beneficial relationship between a fungus and algae or cyanobacteria, where the fungus provides a protective environment for the photosynthetic algae/cyanobacteria, which in turn provides the fungus with food through photosynthesis. Lichens slowly grow outward in a radial pattern, so the more significant the lichen thallus (flat, leaf-like part), the longer it has been growing on the rock surface. The success of lichenometry largely depends on the establishment of growth curves for particular species. It is often criticized for the inability to consider external variables like climate or rock type. A lichen commonly used to date volcanic deposits is *Rhizocarpon geographium.* This seafoam green lichen is found in mountainous regions and has a distinct black outline of fungal hyphae that can be easily identified with the naked eye. The maximum dating range for lichen reaches 10,000 years, but 5,000 is a more practical limit. Nonetheless, this technique is commonly used in polar and alpine regions. It provides a "ballpark" timeline while geologists collect data in the field when complex instrumentation analysis is not yet possible. Moreover, lichens are among the first organisms to colonize volcanic landscapes after a volcanic eruption, as they absorb moisture from the air, can survive extreme temperatures, and are photosynthetic.

What is the role of ice cores in preserving ancient eruptions?

While ice cores can be collected from any glacier, the most extensive compilation of data is derived from those in Greenland and Antarctica, which provide evidence to describe life on the planet from the Pleistocene to the modern day. Ice cores provide valuable archives for studying volcanic eruptions because they can preserve soluble and nonsoluble products of eruptions, including aerosols, carbon dioxide, sulfur, and ash. The aerosols measured in ice cores help determine regional, continental volcanism (e.g., chloride) and volatile metals (e.g., cadmium, bismuth, lead, and thallium). Similarly, carbon dioxide (CO_2) and sulfur (mainly, sulfate and sulfuric acid) are soluble in ice cores and preserve ancient atmospheric levels important to understanding paleoclimate. It is not uncommon that during supervolcano eruptions (Figure 10.2), the levels of carbon dioxide and sulfuric acid on the planet will spike for some duration post-eruption, which has been established through ice core analysis. Lastly, volcanic ash falling on a glacial surface will become included in the ice sheet and preserved as a marker bed. The ash can often be assigned an absolute date through dating techniques previously described, and because the mineralogy of each volcanic eruption is unique, the source can be traced.

Research conducted by Lin et al. (2022) and Plunkett et al. (2023) analyzed volcanic eruptions in Greenland and Antarctica ice cores over the past 60,000 years. Their ice core analysis recorded 1,113 eruptions in Greenland and 740 in Antarctica. Additionally, the researchers found that 25 of the eruptions were more prominent than any volcanic eruption that has happened in the past 2,500 years. Sixty-nine eruptions were more extensive than the 1815 Tambora eruption, ranking 7 on the Volcanic Explosivity Index.

EFFECTS OF VOLCANIC ACTIVITY ON HUMAN LIFE

12

HOW HAS VOLCANIC ACTIVITY INFLUENCED HUMAN LIFE?

Has the human species been affected by supervolcano eruptions?

The eruption of the supervolcano Toba, Northern Sumatra, Indonesia (1,597 m; 2.8191, 98.6077), 74,000 years ago is considered the most significant eruption ever to occur on our planet (Figures 8.4, 10.2). The blast, likely lasting 9 to 14 days, created the deepest crater lake in the world (505 m) and spread volcanic ash 7,000 km away from its vent to cover about 1% of the planet. It has been estimated that this single eruption released 11 billion metric tons of sulfuric acid (H_2SO_4) and 6 million metric tons of sulfur dioxide (SO_2) to instigate a "volcanic global winter" forcing the planet to cool 3.2°C–4.5°C, making the worldwide average just 15°C and contributing to a 1,000-year-long glacial phase. With some exceptions (Ge and Gao, 2003), most interdisciplinary researchers concur that the sudden and dramatic cooling caused by the super eruption of Toba 74,000 years ago resulted in human extinction outside of tropical Africa. The Toba Genetic Bottleneck theory supports a Middle Pleistocene African origin for modern humans (Prothero, 2018; Ambrose, 2003; Gathorne-Hardy and Harcourt-Smith, 2003).

The Toba Genetic Bottleneck theory is a significant concept in understanding the impact of supervolcano eruptions on human history. It describes a period in human evolution where the population drastically decreased, leaving a small surviving population with limited genetic diversity. This led to a narrowing in the human gene pool. Before the eruption, the genus and species of *Homo sapiens* mostly lived in Africa, Asia, and southeastern Europe, not yet reaching Australia or the Americas. With a global population estimated to be around 1 million humans condensed to three continents, research conducted by geneticists and archeologists suggests that post-eruption, only 1,000 to 10,000 breeding pairs of people survived worldwide. Many lines of evidence agree that less than 90% of the human population remained, with similar findings in other large mammals, like pandas, tigers, and chimpanzees (Prothero, 2018). The decrease in population sizes has been linked to the climatic effects of Toba, including drought, crop failure, and famine, confirmed by paleoclimate investigations. Moreover, the genetic bottleneck is more severe in regions closer to the eruption, namely South Asia and Southeast Asia.

What are some examples where archeological discoveries have led to a better understanding of ancient volcanism?

To learn how humans have been affected by volcanic eruptions, geologists often lean on the science of archeology, because it studies ancient human history through the excavation of sites and analysis of artifacts and other physical remains. Taking a multifaceted approach to understanding civilization, archeologists seek to reconstruct the life of entirely vanished societies that included commerce, political organization, religious beliefs, and lifestyle. Archeologists have uncovered whole cities buried in volcanic ash that act as a time capsule for understanding human behavior and response to deadly natural disasters. Notably, the excavation of city centers in Pompeii (40.8230, 24.4288) and Herculaneum (40.8058, 14.3488), Italy, and Akrotiri (36.3583, 25.3969), Greece, revealed clues of the volcanic process (e.g., time of the eruption, rate of deposition) that would not be known through tephrochronology or isotopic dating alone.

In AD 79, Pompeii was a thriving commercial center of 20,000 people, and nearby Herculaneum was a residential town of about 5,000 until the catastrophic eruption of Mount Vesuvius occurred. This eruption completely buried Pompeii, a city of approximately 12 km^2, in 4–6 meters of ash, and Herculaneum, approximately 4 km^2, in 20 meters. These cities would remain undiscovered until the late 16th century, with significant excavations beginning in the mid-18th century. In Pompeii, archeologists discovered an entire metropolis reflective of the Roman Empire, encompassing theaters, shops, private houses, gardens, public baths, swimming pools, and a library. Quickly realizing the magnitude of the disaster, most Pompeiians fled the city center, not wasting time taking valuable possessions like gold jewelry, silver dishware, or bank records with them, which would later become entombed and provide a glimpse of Roman lifestyle. The bodies of approximately 1,000 people were found in the excavation of Pompeii. Volcanic gases likely poisoned the Pompeiians, who did not escape before being buried in ash. Over time, the calcification of ash formed a protective shell around the bodies, allowing organic body matter to decompose slowly, leaving a void representing their anatomical position of death. An Italian archeologist, Giuseppe Fiorelli, first discovered the voids in the 18th century and filled the spaces with plaster to reveal the casts of people in place. Because Herculaneum was buried in three times more ash than Pompeii, most of the town remains unexcavated and enigmatic. However, in the 1980s, 300 skeletons were uncovered near the shore, huddled in boat houses. The preservation of these remains indicates vaporization likely from extremely hot pyroclastic flow deposits.

Also unique to the AD 79 eruption of Mount Vesuvius is a written description of the calamity that provided a detailed sequence of eruptive events. A Roman nobleman, Pliny the Younger, carefully chronicled the eruption of stratovolcano Mount Vesuvius from across the bay and thoroughly documented the event in two letters that he wrote to his mentor, Tacitus. From these letters, it is understood that the eruption began at 1:00 p.m. on August 24 when an explosive eruption column 20 km tall formed above the central vent. The eruption continued for the next 11 hours, with ash and lapilli-sized tephra blown in the predominant wind direction of south-southeast and rained down on Pompeii at a rate of 15 cm per hour. Up until this time, the town of Herculaneum, 6 km upwind, only received a trace of ash during this eruption cycle. At midnight, however, the now gas-depleted ash cloud expanded horizontally instead of vertically, forcing a pyroclastic surge toward Herculaneum, killing all who remained and engulfing the town. A second surge came an hour later, and a third at 5:30 a.m. on August 25 that would entirely engulf Herculaneum. The fourth pyroclastic flow was directed toward Pompeii, already buried in 2 meters of ash, which would shear off roofs and upper stories of buildings and asphyxiate the remaining inhabitants. In his letters, Pliny the Younger provided eyewitness accounts of the generation of the sixth and largest pyroclastic flow at 8:30 a.m. on August 25 that swept across the Bay of Naples. In his honor, volcanic eruptions that are incredibly explosive and produce large amounts of ash are described by volcanologists as having a Plinian eruption style as used on the Volcanic Explosivity Index (VEI) (Figure 8.1). The AD 79 eruption of Mount Vesuvius ranks a VEI 5.0.

The Aegean Island of Thera, commonly referred to today as Santorini is another example of how archeology has aided in understanding volcanism and vice versa. Throughout geologic history, the Thera volcano has had reoccurring cycles of violent Plinian-style eruptions followed by collapse into a 300 m deep sea-filled caldera that later forms a circular chain of small islands that outline its perimeter (Figure 3.3). Around 1600 BCE, a cataclysmic eruption occurred at Thera, which buried the ancient city of Akrotiri in 7 m of volcanic ash. After the partial excavation of the ancient city in 1967, archeologists uncovered approximately 30% or 1 km^2, revealing 40 buildings that represent advancements found in the Bronze Age. A complex urban center built by the Minoans, the structures have sophisticated sewage systems and advanced architecture, including multistory buildings and earthquake-resistant infrastructure.

Unlike Pompeii, there are no clear manuscripts describing the 1600 BCE eruption of Thera, but because human remains have not been found in the ruins of Akrotiri, it is speculated that the Minoans heeded early warning signs that correlate to a precursory veil of ash that dusted the city and subsequent earthquake damage identified within the buried settlement. Following the initial wake-up of Thera, the tephrochronology record provides evidence of four major eruptive phases. The first eruption phase is associated with intense magmatic activity, producing most of the ash burying Akrotiri with estimates between 28 and 41 km^3 dense-rock equivalent (DRE)—the eruption's intensity mounted during the second and third phases with pyroclastic surges and significant lava fountaining. The final phase concluded with lava flows and lahars, and ultimately, another caldera collapse generated a megatsunami wave between 35 and 150 m tall, erasing the northern coastline of Crete approximately 110 km away.

In archeology, the 1600 BCE eruption of Thera is called the Minoan eruption. The ash deposition from the first eruptive phase is an important marker bed used throughout the entire Eastern Mediterranean to identify the absolute age of artifacts. However, controversy exists around the exact timing of the event with radiocarbon dates from multiple objects and sites ranging between 1605 and 1675 BCE, from tree rings ranging between 1524 and 1681 BCE, and isotopic dates retrieved from ash in Greenland ice cores ranging from 1555 to 1654 BCE. The error margin of the dating techniques, spanning a century, captures the discussion as to whether the eruption of Thera caused the Minoan civilization to collapse in 1450 BCE. Nonetheless, this eruption, 100 times more powerful than the AD 79 eruption of Mount Vesuvius, was among the most energetic explosions in human history, ranking 7.0 on the Volcanic Explosivity Index (Figure 8.1).

Could volcanic activity institute war?

While understanding the effects of volcanism is complex, some researchers link the start of the French Revolution on May 5, 1789, to a volcanic eruption in Iceland. But why? Prior to the French Revolution, some factors created turmoil in France. First, the country was experiencing financial difficulties as King Louis XVI wanted to increase taxes to pay off his court's debts. The working class paid the vast majority of the country's taxes, causing them to become increasingly irritated with the king and elite socialites. Next, popular French philosophers, namely Voltaire and Rousseau, supported the working class by promoting new ideas based on equality, liberty, and the separation of church and state. Lastly, relating to volcanics, the climate of France turned dry and cold, leading to strained harvests, causing food prices to soar and famine to spread throughout France and Europe. Food scarcity was the final indignity for the working class that would initiate the revolution.

Some 2,000 km northwest of France and 6 years prior, a fissure eruption began in Iceland on June 8, 1783. A fissure eruption occurs when linear underground passageways, called dikes, get filled with magma. This effusive eruption, the Laki Fires (64.0780, –18.2383), only lasted 8 months. However, a 24-km-long fissure allowed 42 billion metric tons of basaltic lava to quickly arrive at the surface, making this event Iceland's second-largest flow in historical time. Throughout 10 episodes, 130 vents would form, releasing 120 million metric tons total sulfuric acid aerosol into the atmosphere and deadly amounts of fluoride. In Iceland, the Laki Fires are the island's worst natural disaster, as one-fifth of the population perished by 1785 from SO_2 inhalation, fluorosis, or famine (Thordarson and Self, 2003).

Following the predominant wind direction, the sulfuric acid emissions from the Laki Fires would be carried in the jet stream toward Europe, forming a visible, dry fog that smelled of rotten eggs and would rain down to contaminate the landscape. While Europeans were unaware of the source of the well-documented "Acid Haze," American diplomat and scientist Benjamin Franklin served as the American ambassador to France and made the connection. Historical accounts and data collected from varying climate proxies concur that summer 1783 and several years to follow were characterized by extreme and unusual weather, including exceptionally hot summers in western Europe followed by a decrease in winter temperatures by 1.3°C, extending and exacerbating cold winter temperatures and thereby reducing growing seasons. The famine that resulted in France prior to the French Revolution could have been a direct effect of the Laki Fires.

What are some examples of cultures incorporating religious and indigenous beliefs into their practices honoring volcanoes?

Many cultures have revered volcanoes as landscape features connecting nature to powerful creators, inspiring the religious worship of gods and goddesses, and influencing mythology. Volcanoes, representing the elemental characteristics of Earth, wind, and fire, have been portrayed as destructive and constructive forces. Oral tradition, art, and the written word, passed on through generations, often alter specifics but keep the continuum of the moral message intact. In many volcanic regions, it remains very important to share ancient stories to respect ancestors, culture, and nature; Hawaii, Mexico, and Indonesia are just a few examples.

Oral tradition speaks of the Hawaiian Goddess called Pelé. This myth originates around the 15th century and describes how Pelé first arrived on the Big Island of Hawaii in a canoe from Tahiti. The story says that Pelé was banished from her homeland by her father for her temper and for seducing her sister's husband. While Pelé would try to create new volcanoes, her wrought sister, Namako Kaha'l, would intentionally flood them with her powers. Eventually, Pelé found refuge in the crater Halema'uma'u at the top of Kilauea, where she is the goddess of volcanoes, fire, lightning, wind, and dance. She is known as the creator of the Hawaiian Islands and is embodied by the lava and other processes associated with volcanics. Perhaps most importantly, it is customary in Hawaiian tradition to ask Pelé for permission to travel through her land and sacred landscape. Additionally, legend has it that Pelé will enact a curse on anyone who collects rocks from the island as they are sacred, representing her children.

In Mexico, the Aztecs also share a story from the 15th century that links two stratovolcanoes: Popocatepetl (1,519 m; 19.0322. –98.6261) and Iztaccihuatl (5,230 m; 19.188, –98.6422). These volcanoes that tower over Mexico City are worshipped as deities and are linked to an enchanting love story. Popocatepetl represents a valiant and victorious warrior away from his beloved, Princess Iztaccihuatl, during a long and drawn-out battle. Word was sent to the princess from the battlefield, telling her that Popocatepetl was killed. In grief, she laid down and died. When Popocatepetl returned home to find the princess dead, he took up a torch and knelt before his beloved, forever guarding her eternal slumber, and for this, he built great mountains. On one, he buried the body of Princess Iztaccihuatl, and on the other (Popocatepetl), he stands eternally, holding her funeral torch in memoriam. Therefore, from that moment onward, Popocatepetl and Iztaccihuatl would remain together, forever facing one another. The legend continues that whenever Popocatepetl yearns for his companion, his heart ignites in embers, and his torch begins to smoke. Even today, Popocatepetl remains an active volcano.

Instead of happy endings, some ancient tales focus on appeasing capricious gods at dangerous and unpredictable volcanoes. Two examples come from Nicaragua and Indonesia. As did Aztecs, Mayans, and Incas, indigenous

Nicaraguans long believed that volcanoes would only stay quiet if human sacrifices occurred. Specifically, the legend of Volcano Consigunia (4,613 m; 12.9847, –87.5641) required that a young child be thrown into the crater every 25 years, and at Volcano Masaya (256 m; 11.9725, –86.0880), a young woman. Likewise, for centuries in the Hindu religion, it was customary to sacrifice humans at Mount Bromo (2,329 m; –7.941, 112.9527) on the island of Java in Indonesia. This legend stems from a couple named Roro Anteng and Joko Seger. This couple remained childless after many years of marriage until, one day, they meditated on top of Mount Bromo and asked the mountain gods to bless the children. The gods agreed by granting Roro and Joko 24 children under the condition that the 25th child would be thrown into the volcano as a human sacrifice. The tradition of the sacrifice continues today, but instead of humans being offered, the locals toss chickens, goats, and vegetables into the caldera during a festival called Yadnya Kasada.

How have volcanoes been portrayed in Greek and Roman mythology?

In Greek and Roman mythology, volcanoes were often linked to gods and dramatic stories involving the Earth's divine strength. The gods were portrayed as expressions of anger, power, or the outcome of celestial battles. Some famous myths include stories of the Greek god of fire Hephaestus (or his Roman counterpart Vulcan) and the Battle of the Titans.

Hephaestus (Vulcan) is described as the son of Zeus (Jupiter) and Hera (Juno), the husband of Aphrodite (Venus), and one of twelve Olympians. In mythology, Hephaestus and Vulcan represent divine anger and intense energy that originates in the depths of the Earth. Additionally, the fire gods symbolize metal fabrication, with several images linking Hephaestus and Vulcan to manufacturing thunderbolts for Zeus. Their workshops are often depicted underneath Mount Etna (3,269 m; 37.7513, 14.9933), Italy. Likewise, the Greek giant Typhon, a creature associated with chaos and fire, is said to be imprisoned beneath Mount Etna, and his rage is expressed through volcanic eruptions.

In Greek mythology, the Titanomachy, or Battle of the Titans, was a war between the Titans and the Olympian gods. This mythology, tied to volcanology, stems from the Titans being perceived as huge man-creatures born of the Earth to attack gods. The Titans hid underground until the war, when, like a volcano, they erupted to the surface with an unstoppable force.

Did the eruption of Ilopango initiate the decline of the Mayan civilization?

The eruption of Ilopango (639 m; 13.7019, –89.1047), El Salvador, in AD 536, was one of the largest volcanic eruptions to occur in Central America during the Holocene. Ranking a Volcanic Explosivity Index (VEI) of 7.0, this caldera-forming Plinian-eruption forced the globe into a volcanic winter through its exuberant aerosol emissions, dropping temperatures as much as 2.5°C globally. The cooling trend, referred to as the Late Antique Little Ice Age, would last for 24 years, from AD 536 to 560. Locally, 0.5 cm of ash blanketed 2 million km^2 of Central America while making 20,000 km^2 nearest Ilopango (including areas of Guatemala and Honduras) uninhabitable, with ash accumulation greater than 35 cm (Dull et al., 2019).

From a plethora of archaeological evidence, it has been found that before the eruption, the area around Ilopango was a densely populated (40 people/km^2) and intensely cultivated region inhabited by Mayans. The paroxysmal eruption proved to be devastating, as a 2,000 km^2 area adjacent to the vent was inundated by a 3-m-thick pyroclastic density current that took the lives of ~80,000 people (Dull et al., 2019). With unlivable conditions left in the periphery of Ilopongo, it is estimated that 400,000 Mayans were forced to flee the region and settle elsewhere, causing large-scale demographic collapse following the event. Additionally, the initial decline of the Mayan civilization is attributed to post-eruption disease outbreaks, likely resulting from the inability to reach food or clean water soon enough.

Although it is hard to know where the survivors settled, it has been postulated based on ceramic affinities that many fled to the north to major Mayan centers like Kaminaljuyu and Copan (Dull et al., 2001). The authors also suggest that the greatly affected regions of El Salvador and southeastern Guatemala did not completely recover until the 7th century. Eventually, the Mayan civilization rebounded from the eruption. However, the catastrophe may have reshaped the economic, political, and social landscape leading up to the Classic Mayan collapse in the 9th century.

What is the oldest known description of volcanic activity?

It is no surprise that many archeological discoveries display artifacts depicting human interaction with volcanoes over historic time frames. Although pre-humans did not have the capacity to understand the science behind these landform features, artistic renditions found on cave walls, pots, textiles, and jewelry clearly show that volcanoes were awe-inspiring, active, and important structures in their natural environment.

To date, the oldest known description of volcanic activity was found at Chauvet-Pont d'Arc Cave, France (44.3877, 4.4158), dating back 36,000 years. This period coincides with the arrival in western Europe of anatomically modern humans (*Homo sapiens*) associated with the Aurignacian culture. Through a series of red fan-shaped bursts in a radiating pattern coming out of a triangle, the cave drawings are interpreted to reveal an explosive volcanic eruption that resembles lava fountain activity. Nomade et al. (2016) explored the Bas-Vivarais volcanic field 35 km northwest of the cave to confirm this assumption. The Ar40/Ar39 ages acquired from the ash collected at Bas-Vivarais overlapped the radiocarbon and thermoluminescence ages of the artifacts found within the cave. Bas-Vivarais volcanism is characterized by phreatomagmatic eruptions that produce maars that precede or follow Strombolian activity and are associated with effusive lava flows.

Prior to the discovery of Chauvet-Pont d'Arc Cave, a Neolithic mural found in the Catalhoyuk Cave, Turkey (37.1975, 35.8152), was thought to be the oldest evidence of human connection to volcanic activity. The 8,000-year-old mural illustrates an explosive summit eruption of the stratovolcano Mount Hasan, located ~130 km northeast of the cave entrance. Red and black pigments emphasize the dramatic volcanic activity, possibly the formation of glowing lava or ash clouds. Similar to Chauvet-Pont d'Arc Cave, geological evidence substantiates such eruption type. For perspective, this mural is 6 millennia older than a fresco uncovered at Pompeii that illustrates the god of wine, Bacchus, presiding over vineyards near Mount Vesuvius, Italy.

What are the effects of volcanic events on water supply?

Volcanic eruptions can have short-term and long-term effects on water supply that can impact its quality and availability, causing disruptions to ecological and human life. Contamination of the water supply can occur in various forms. Particulate matter falling as ash can enter rivers, lakes, and reservoirs, making the water undrinkable by influencing the water's acidity and turbidity. While filters can help remove these attributes from the water, they are often not readily accessible post-eruption or are too small to keep up with population demand on the water supply, as was experienced during the 2022 Hunga Tonga eruption. Next, large quantities of ash, or volcanic debris, can mix with rain or melting ice to form lahars that bury and block rivers, disrupting drainage systems and carrying debris into drinking water sources. Lahars can completely alter a river course to change the area's hydrology permanently. Dramatically, this was seen during the Mount St. Helens 1980 eruption, when a massive lahar filled the North

Fork of the Toutle River with sediment, raising the riverbed over 46 m to alter the course of the river permanently. Lastly, eruptions releasing volcanic gases containing sulfur dioxide (SO_2), fluoride (F^-), or soluble heavy metals (e.g., lead, cadmium, arsenic, copper, mercury) can mix with atmospheric moisture and fall back to Earth, contaminating the groundwater or surface water with hazardous chemicals. Drinking water containing elevated levels of fluoride, for example, can lead to a deadly condition called fluorosis, affecting bones and teeth. Poisoning from fluorosis was responsible for the death of one-fifth of Iceland's population during the Laki Fires fissure eruption of 1783. Likewise, even the smallest amount of arsenic or mercury found in a water supply has fatal consequences.

PLANETARY GEOLOGY

13

PLANETARY VOLCANISM

What are planetary geology and planetary volcanism?

Planetary geology is the study of celestial solid bodies of matter moving in an elliptical orbit around a central star. Specifically, this field of science explores how these bodies are formed and changed over time. Our Solar System consists of eight planets: Mercury, Venus, Earth, Mars, Jupiter, Saturn, Uranus, and Neptune (Figure 13.1). The planets closest to the Sun (Mercury, Venus, and Earth) are referred to as inner planets and are composed mostly of rock and metal, whereas the remaining outer planets have greater proportions of hydrogen, helium, and other gases. A planet orbits a star, also called the Sun, while a moon orbits a planet. Each planet has its own gravitational field that locks their moon, or moons, within their orbit. All moons in the Solar System rotate on an axis and many moons, including the Earth's moon, are tidally locked to their planet, meaning they rotate at the same rate as they orbit. This is why we only see the near side of the Moon. Within the Solar System there is rocky debris such as meteorites or asteroids, and comets made of ice and dust. On Earth, this debris usually burns up in the upper atmosphere before colliding with the surface.

Planetary volcanism focuses on the surface of planets to determine if eruptions of magma take, or have taken, place in the

Figure 13.1 Diagram of the Solar System.

past. This field of science is interested in determining if heat or gases emitted during planetary volcanic eruptions shape their surface or atmosphere. Scientific information is gathered via land-based telescopes, spacecraft, satellites, and tools that measure heat and radiation.

Is the Moon differentiated like the Earth, and is it tectonically active?

Similar to the Earth, the Moon is differentiated into sections including a solid inner core, a solid-state submolten mantle, and a solid crust (Figures 1.1 and 13.2). The Moon's inner core is likely made up of nickel, iron, and sulfur and is proportionally smaller to its overall size (radius approximately 240 km) constituting about 15% of its diameter. The high-density, mafic, lunar mantle is iron- and magnesium-rich and occupies the majority of the Moon's constitution. The exterior surface of the Moon is wildly variable in thickness depending on its location, namely the near side (the side facing the Earth) or the far side. Because the Moon is tidally locked to the Earth, our nightly observation only reflects the Moon's near side. The asymmetrical distribution of thickness may be the result of uneven heating during its formation 4.3 BYA, with the far side having thicker crust compared to the near side due to the intense heat radiating from the young Earth while the Moon was solidifying. This phenomenon, referred to as lunar dichotomy, allowed the near side crust to partially melt and thin.

The Moon is not tectonically active for several reasons. Unlike the Earth, which has a partially molten asthenosphere that promotes plate tectonic movement (Figures 1.1 and 1.3), the Moon's mantle is considered rigid. The lack of pliability is due to the Moon's mantle not containing water, and also because its inner core radiates little heat, not allowing for the production

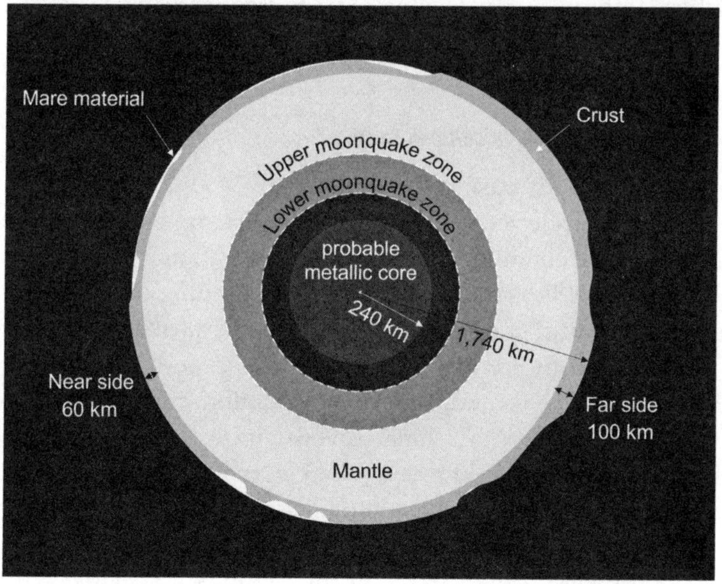

Figure 13.2 A cross section of the Moon's interior showing locations of moonquake zones and thickness variations between near side and far side crust (not to scale).

of fluid magma necessary for convection cell formation. While the Moon lacks convergent and divergent plate boundaries (Figures 1.5 and 1.6), it does experience "moonquakes," or releases of seismic energy similar to earthquakes. The moonquakes result from tidal stresses, thermal expansion and contraction, internal cooling, and impact. Release of energy occurs from the influence of tides at a depth around 700 km. The tidal forces from the Earth's gravity pulls on the Moon and, as internal stresses builds up, they are released as quakes. Next, thermal moonquakes occur due to extreme shifts in daytime versus nighttime surface temperatures, ranging between 127°C and −173°C. As the Moon's crust expands and contracts, it can fracture, causing seismic activity. Over geologic time, the Moon's inner core has continued to cool, causing its overall size to shrink. This contraction releases shallow (50–150 km) moonquakes that can reach a magnitude of 5.5 on the Richter scale. And lastly, because the Moon does not have an atmosphere to protect it, collisional objects frequently slam into its surface, dispersing shockwaves.

Is the Earth's moon volcanically active?

The presence of past volcanism on the Moon is understood from data collected by Apollo missions organized by the National Aeronautics and Space Administration (NASA) from 1969 to 1972. Beyond the importance of landing on the Moon itself, scientific missions focused on the collection of lunar samples and the conduction of scientific experiments on the Moon's surface to understand its formation, mineral composition, and age. With a lunar rover vehicle (LRV), a battery-powered four-wheeled rover, 382 kg of material was collected from the Moon from the last three missions of the Apollo program (15, 16, 17) that geologists continue to analyze today.

Volcanism on the surface of the Moon is associated with its formation, approximately 4.46 BYA, when a large object, possibly the size of Mars, hit the Earth. This event is known as the Theia Impact. This massive collision created debris that would eventually combine, cool, solidify, and be trapped by the Earth's gravitational pull. Initially, the newly formed sphere was likely covered in a molten ocean of magma composed of anorthosite, a mineral that is rich in plagioclase feldspar (although not silica-rich enough to be considered felsic), while the denser, iron-rich minerals like olivine and pyroxene would sink to the bottom (Figures 2.3 and 2.4). Thus, the light-colored areas of the Moon, known as the Highlands, are exposures of original anorthositic lunar crust. And the dark features visible with the naked-eye from Earth (i.e., "The Man in the Moon") are younger (4.2–1.2 BY) impact basins, created from collisional asteroids and comets, that were infilled with basaltic lava. These low-lying areas, referred to as dark maria (plural) or mare (singular), represent ancient lava flows that early astronomers mistakenly believed to be water bodies, hence the Latin naming that translates to "seas." Lunar mare basalts (LMB), found in the dark maria, cover approximately 20% of the Moon's surface and are more common on the near side of the Moon. Like terrestrial basalts (Figure 2.6 C), LMB are made of pyroxene and plagioclase feldspar (Figure 2.3) formed by partial melting of the Moon's mantle (Figure 13.1). Additionally, as on Earth, it is speculated that the small grain size (>1 mm) indicates rapid cooling and that vesicles indicate the escape of gases during the eruption.

As expected with the deposition of LMB, and as identified in the low-slope plains of the dark maria, the Moon's volcanism is described as being predominately effusive for a variety of reasons. First, the chemical composition of the Moon's crust and mantle is mostly mafic, lending itself to magma that has low viscosity, or low resistance to flow. While this has been confirmed by geochemical analysis, examination of high-resolution imagery of the Moon sent from a satellite (Lunar Reconnaissance Orbiter) since 2009 has provided evidence for geologic structures associated with effusive volcanism including pahoehoe (Mare Serenitatis) (Figure 4.2), lobe fronts (Mare Tranquillitatis), and lava channels and tubes (Mare Imbrium).

Next, the thin consistency of the lava is exacerbated by the weak gravitational field on the Moon, which allows it to spread easily, forming maria that span hundreds to thousands of kilometers in length and width; this is comparable to large flood basalt deposits found on Earth (e.g., Siberian Lava Traps, Russia; Deccan Lava Traps, India; Columbia River Basalts, Washington). Moreover, the Moon experiences nonexplosive volcanism because the magma has a very low volatile content due to its low gravity field, and the extreme heat released during formation caused water and gases to escape into space. And, lastly, because the Moon lacks an atmosphere, the accumulation of gasses does not commonly occur, preventing explosions from happening.

Despite the recognition that the craters seen on the Moon are formed from material impact (i.e., asteroids, meteorites, comets), and not from caldera collapse, there is some evidence of other volcanic processes that have taken place. Features that have the appearance of broad sloping shield volcanoes, lava domes, and central-vent cones are clustered in fields within the maria (Figure 3.2). One of these complexes, called Marius Hills, is several hundred meters high. Moreover, Apollo 15 and 17 missions returned to Earth with samples of green and orange volcanic glass that have the appearance of obsidian (Figure 2.6 E). Although early speculation suggested that the glass was melt from impact craters, geochemical analysis revealed that the glass spheres represent the products of a fine spray of low-viscosity lava into space that is equated to lava fountains commonly seen in Hawaii or Iceland. The Apollo 15 green glass is very rich in magnesium and extremely low in titanium, whereas the Apollo 17 orange glass is very rich in titanium. The surface of both glasses is coated with a variety of volatile elements including lead, zinc, and halogens such as chlorine.

Prior to a recent discovery in 2024, it was thought that the last volcanic activity on the Moon occurred 1.2 BYA, with peak magmatism between 3.8 and 3.1 BYA. Isotopic analysis of lunar material delivered to Earth by China's Chang'e 5 spacecraft revealed ages of volcanically generated material as young as 100 MY (Wang et al., 2024). The authors concluded that because the lunar material contained high levels of heat-generating elements (e.g., thorium and other rare earth minerals), it could cause melting in the lunar mantle and initiate an eruption. The location where this lunar material was collected is referred to as an irregular mare patch (IMP). An IMP is an ideal location for young rock exploration because these areas have experienced less material bombardment than surrounding terrains, like lunar highlands or maria. Additionally, because the lunar surface is younger in IMRs, these areas have experienced less erosion and are easier to identify in satellite imagery. At this time, geologists do not have any indication that there are currently active volcanoes on the Moon, nor do they anticipate another eruption soon. Instead, it is likely that the Moon's mantle has cooled to the point that magma can no longer be generated.

Besides the Earth, what other planets, or moons, are currently volcanically active?

Io, one of Jupiter's four largest moons, is the only orbital body outside of the Earth known to have large-scale active volcanism today. Io, slightly larger than the Earth's Moon, is similarly differentiated into parts including an inner iron core, silica-rich mantle, and a thin outer crust dominated by sulfur and silicate materials. However, about 20% of Io's mantle is likely molten magma. The magma on Io originates from a gravitational tug-of-war between Jupiter and its other neighboring moons (Europa and Ganymede). This constant stretching and squeezing, called tidal flexing, generates intense frictional heat inside Io. As on Earth, the magma rises through cracks and fractures in Io's lithosphere, erupting at the surface. Io's ultramafic, magnesium-rich lava exceeds temperatures of 2000 K. This is particularly interesting to geologists because ultramafic lavas have rarely been erupted on Earth since the Proterozoic Eon, or 1.8 BYA.

At its closest distance, Io is approximately 588 million km away from the Earth and was first discovered by Galileo in 1610. As the advancement of land-based telescopes continued through the 1900s, astronomers could begin to discern a mosaic of colors on Io's surface that were surprisingly similar to those found in geothermal fields on planet Earth. Researchers were left with the question: Could the yellows, oranges, and reds indicate sulfur emissions on Io? In 1972, the Pioneer 1 spacecraft flew past Io, relaying images for the first time that would confirm the suspicions that Io was gaseous and volcanically active (Figure 13.3). As spacecraft imagery improved with Voyagers 1 and 2, geologists identified more than 400 volcanoes on Io's surface with volcanic plumes reaching 300 km above its surface. Additionally, with Voyager spacecraft taking images in two consecutive years (1972 and 1973) landscape comparisons

Spacecraft Name	Year of Flyby	Purpose
Pioneer 10 and Pioneer 11	1972 and 1973	These were the first spacecraft to study the Jupiter system in detail. Pioneer 11 was the first to take close measurements of Io and confirmed its active volcanism.
Voyager 1 and Voyager 2	1979	In 1979, these spacecrafts captured images of Io, including evidence of volcanic eruptions.
Galileo	1995 to 2002	Between 1995 and 2002, this spacecraft made multiple flybys of Io, providing the closest views of the moon to date.
Cassini	2000	This spacecraft studied Io while enroute to Saturn
New Horizons	2007	This spacecraft flew past Jupiter to gain gravity assist on its way to Pluto, passing Io along the way.
Juno	2016-present	This spacecraft has been orbiting Jupiter since 2016 and has made multiple flybys of Io.

Figure 13.3 Table listing spacecraft name, year of flyby, and purpose of flight relating to the investigations of Io.

could be made. In doing so, geologists determined that the Io's surface dramatically changes over short time scales, as previously formed calderas having diameters 20–200 km in diameter were infilled with magma over the course of a year.

Io is estimated to have formed 4.5 Byo, roughly the same time as Jupiter and the rest of the Solar System. However, the entire surface of Io is suspected to be extremely young in geological terms as it is constantly reshaped by intense volcanic activity. As an example, in 2024 the Juno spacecraft detected a massive hot spot in Io's southern hemisphere the size of Earth's Lake Superior with eruptions emitting 6X the total energy of all Earth's power plants (NASA, 2025). Moreover, the volcanism on Io has been likened to that of steady-state magma generation seen in long-term eruptions on Earth such as Arenal (Costa Rica) or Stromboli (Italy).

What is the largest volcano in the Solar System?

The largest volcano in the Solar System is found on the surface of Mars. A shield volcano (Figure 3.2), Olympus Mons has a base of 600 km in diameter and reaches 25 km above the surface. For perspective, Olympus Mons is approximately the size of Arizona, could fit within the borders of France, and is almost three times the height of the tallest peak on Earth, Mount Everest (elevation 8.8 km). The volcanic structure is so large that it was first identified by a land-based telescope in 1900s and later confirmed in 1971 by Mariner 9, the first spacecraft to take images of the Martian landscape.

There is no direct evidence that Martian eruptive events are somehow different from volcanism seen on Earth. The composition of Mars has largely been studied through three different sources: spacecraft and Earth-based telescopes, land rover data collection, and a special group of meteorites (Shergottie, Nakhlite, and Chassigny) for which there is compelling evidence that these rocks were blasted from the surface of Mars and eventually fell to Earth. These independent lines of reasoning conclude that the Martian landscape is about twice as rich in iron than Earth, hence the planet's distinctive red color, and that the outer crust of Mars is primarily basaltic. In fact, the composition and structure of Olympus Mons are often compared to the Hawaiian island, Mauna Loa (19.4723, −155.5922).

Despite having many volcanic similarities to our planet, the question still remains: How did Mars, a planet half the size of Earth, generate a volcano the size of Olympus Mons? Lacking any obvious signs of plate tectonics, it is speculated that Olympus Mons resulted from mantle sources that released magma from its lithosphere. That is, the enormity of this volcano was created by a hot spot. Different from planet Earth, however, the hot spot and the crust remain stationary, allowing the magma to pile in one location forming a giant volcano, rather than a series of small volcanic islands (Figures 1.7 and 1.8). Additionally, extremely high eruption rates may have been encouraged by the magma's low viscosity, which is supported by the low gravity field found on Mars's surface.

Through a variety of dating techniques, it has been found that Olympus Mons first formed during a period of intense volcanism on Mars occurring 3.5 BYA. The complex summit of Olympus Mons reveals a caldera that has gone through multiple episodes of collapse (Figure 3.3), with the last event happening 25 million years ago. The youngest lava flows are estimated to be 2 million years old. This estimate comes from a unique dating technique called crater counting. Here, the number and size of impact craters on different lava flows help estimate relative ages; older surfaces accumulate more craters, while younger flows have fewer. While Mars has experienced significant volcanic activity in the past, there is no definitive evidence of similar occurrences happening today.

What planet has the most volcanoes?

With 1,600 dormant volcanoes, the planet Venus has the most of any in our Solar System. This total, which represents only those volcanoes that are well-preserved and highly recognizable by imagery, exceeds the amount found on Earth and Mars combined. Venus is often compared to Earth because of its close proximity (41 million km away), and its similarities in planetary size (diameter, 12,103 km), density (5.24 g/cm^3), and composition (mafic). While Venus is thought to be differentiated into parts like the Earth, it does not have plate tectonics to aid in recycling carbon dioxide that is expelled into the adjacent atmosphere. For this reason, Venus's atmosphere is primarily composed of carbon dioxide that traps heat through a runaway greenhouse effect forcing the planet's surface temperature to elevate to extreme highs (~475°C). Additionally, while Venus likely had water in the past, its hot surface conditions caused it to evaporated from the surface. A consequence of having an atmosphere with high levels of CO_2 and no water present on its surface means that little erosion can take place. Without erosion, volcano formations can remain in pristine condition, allowing spacecraft to photograph their structures despite the majority being 2.5 BYO (Bindschadler, 1995).

Where is cryovolcanism found, and what is it?

Beyond the orbital reaches of Mars, frigid temperatures of the Solar System lend themselves to the formation of ice-rich bodies that are found as satellites (moons) of large planets (Jupiter—Europa, Ganymede; Saturn—Enceladus, Titan; Neptune—Triton) or as individual bodies orbiting the sun, as is the case for the dwarf planet Pluto. These icy bodies contain water and other compounds that were too volatile to remain in the warmer, interior portions of the Solar System closer to the Sun. Cryovolcanism, or ice volcanism, is the eruption of liquid or vapor phases of water or other volatiles (e.g., ammonia, carbon monoxide, methane) that would be frozen solid at the normal temperature of the icy satellite's surface.

Satellites experiencing cryovolcanism are considered differentiated into parts likely having a solid inner core, but in place of a mantle there is a "global ocean" that exists underneath the icy crust. Two spacecraft-explored satellites that are speculated to have such features include Europa, one of Jupiter's four largest moons, and Saturn's sixth largest moon, Enceladus. Geologists have evidence to presume that, approximately 150 km below their crust, where temperatures are around −160°C, they contain bodies of water larger than all of the oceans on Earth combined. Similar to Earth processes, ocean water found on these satellites is salty because minerals from their rocky mantles become dissolved into solution through heat supplied by radioactive decay. Because the ocean water is under great pressure on Europa and Enceladus, it reaches the surface through cracks in the ice, where it quickly freezes at the surface. This hydrothermal venting is thought to be equivalent to the creation of black smokers found at divergent plate boundaries on Earth (Figure 1.5).

What are the types of cryovolcanism?

Akin to eruption styles witnessed on Earth, cryovolcanism can result in both explosive and nonexplosive events. Explosive cryovolcanic eruptions are expected to be initiated by the exsolution of dissolved volatile gases as pressure drops with the ascension of the cryomagma, mimicking mechanisms of explosive volcanism on terrestrial planets (Figure 1.3). However, explosive cryovolcanism differs from volcanism on Earth because instead of being propelled by gaseous water, carbon dioxide, and sulfur dioxide, it is more likely to be driven by methane and carbon monoxide. As a result, some cryovolcanic landscapes, like those of Neptune's moon Triton, are similar to many silicate volcanic terrains, including what appear to be volcanic rifts, calderas, cinder cones, stratovolcanoes, and domes (Kargel, 1994). Although it has been noted that cryoclastic material is formed during violent eruptions, the debris is often immediately covered by frozen gases that cover the newly pulverized material (Geissler, 2015). Images of Triton have also revealed flow deposits that are hundreds to thousands of meters thick, implying highly viscous lavas resembling those that are felsic on planet Earth. More similar to mafic terrestrial eruptions, effusive, or gentle eruptions are characterized by widespread cryolava flows covering preexisting landscapes, as on Jovian satellite in Saturn's orbit. These cryoflows form thin plains, similar in structure to flood basalts or shield volcanoes, that imply that they have a low viscosity.

There are three conditions that must be met for either explosive or nonexplosive cryovolcanism to occur. First, there must be sufficient production of cryomagma held within a reservoir. The cryomagma, a slurry of water and volatiles, is made from the dissolution of the rocky mantle by heat that can be generated by radioactive decay of isotopes or by tidal fluxing, gravitational interactions with a parent planet (e.g., Saturn's effect on Enceladus) to produce internal friction. Next, because there are no plate tectonic forces or convection cells to encourage the ascent of the magma to the satellite's surface, the buoyancy of the cryomagma is dependent on its water content. The uplift of the cryomagma is driven to the surface by water that expands upon freezing, allowing it to becomes lighter in its solid form than when in its liquid phase. Lastly, the cryomagma must be allowed to escape at the satellite's surface. This process occurs when subsurface pressure is released forming fissures in the icy crust.

Why is it important to study planetary volcanism?

Major advancements in planetary volcanism have occurred over the last 70 years with the reception of information from spacecraft launched decades ago and continued improvements to technology. As more data is analyzed, it has become understood that planetary volcanism is important for several reasons that could impact humanity. Fundamentally, the science of geology is grounded in Jim Hutton's ideology that "The present is a key to the past." Because volcanic activity shapes extraterrestrial landscapes that can be made visible by varied instrumentation, planetary volcanism provides a glimpse of the ancient formation of the Solar System. At its simplest, the surface exposure of rock provides insights into a planet's internal heat, composition, and history, allowing geologists to compare and contrast Earth's crust and atmosphere to better understand our evolution. Moreover, further exploration of the temperature and chemical properties in the global ocean layer of cryoplanets could reveal parallels to those that prevailed in Earth's oceans over 3.5 BYA, providing habitat for the first signatures of life to establish (i.e., cyanobacteria). Thus, Europa Clipper, a NASA spacecraft launched on October 14, 2024, is anticipated to conduct a detailed study of Jupiter's moon starting in April 2030 that may yield scientific evidence that life exists 2.9 billion km away from Earth. This focused data collection will be in addition to understanding Europa's surface and atmospheric interactions. From monitoring volcanic eruptions that release greenhouse gases like carbon dioxide, sulfur dioxide, and water vapor could come invaluable information as our planet continues to warm. For example, maybe our nearest neighbor, Venus, with its thick CO_2-rich atmosphere, will help us to understand how volcanic activity contributes to runaway global warming and how to assess potential hazards of future super volcano eruptions on Earth. Additionally, as the Earth's Moon and Mars are on the forefront of interest, there exists the potential that extraterrestrial landscapes will provide future resources and habitat for humans.

GLOSSARY

A'a—A type of basaltic lava characterized by a rough, clinkery, and jagged surface texture, typically produced during Hawaiian-style eruptions.

Absolute Time—The determination of specific numerical ages for geological materials, often using radiometric dating techniques.

Aerosol—Tiny solid or liquid particles suspended in the atmosphere, often produced by volcanic eruptions, which can affect climate by reflecting sunlight.

Albedo Effect—The measure of a surface's reflectivity; for example, light-colored surfaces such as ice reflect more sunlight, influencing local and global temperatures.

Amplifying Feedback Cycle—A process where an initial change triggers effects that further enhance that change; for instance, melting ice reducing reflectivity and causing additional warming.

Asthenosphere—The relatively soft, ductile portion of the upper mantle beneath the lithosphere that permits the movement of tectonic plates.

Basalt—A common mafic extrusive igneous rock formed by the rapid cooling of basaltic lava at the surface, typically dark in color and low in silica.

Bathymetry—The measurement and study of underwater topography, including ocean depths and seafloor features.

Black Smoker—A type of hydrothermal vent on the seafloor that emits superheated, mineral-rich water, forming chimney-like structures and supporting unique ecosystems.

Bowen's Reaction Series—A model that describes the sequential crystallization of minerals from cooling magma, indicating which minerals form at specific temperature ranges.

Caldera—A large, basin-like depression formed when a volcano collapses after a massive eruption empties its magma chamber.

Cinder Cone—A small, steep-sided volcano built from the accumulation of volcanic ash, cinders, and other pyroclastic fragments from explosive eruptions.

Country Rock—The preexisting rock before undergoing intrusions.

Debris Flow—A fast-moving, gravity-driven mass of loose rock, soil, and water that travels down slopes, often triggered by heavy rainfall, volcanic eruptions, or earthquakes.

Dense-Rock Equivalent (DRE)—A method for estimating the volume of erupted volcanic material by converting fragmented deposits into the equivalent volume of solid rock.

Diatreme—A volcanic pipe formed by explosive interactions between rising magma and groundwater, often resulting in a carrot-shaped, brecciated structure.

Dike—A tabular intrusion of igneous rock that cuts across preexisting rock layers when magma forces its way through fractures.

Divergent Plate Boundary—A tectonic boundary where two plates move apart, allowing magma to rise and create new crust, as seen at mid-ocean ridges.

Earthquake Swarm—A sequence of numerous earthquakes occurring in a localized area over a short period without one dominant mainshock.

Extrusive Igneous Rock—Rock formed when magma reaches the surface and cools rapidly, typically exhibiting a fine grained texture.

Feldspar—A group of abundant rock-forming minerals found in Earth's crust, essential components of many igneous, metamorphic, and sedimentary rocks.

Felsic—Describing igneous rocks or magma that are rich in light-colored minerals (like quartz and feldspar) and have a high silica content.

Fissure Eruption—A volcanic eruption that occurs along a linear crack or fracture in the Earth's surface rather than from a single, centralized vent.

Flood Basalt—A large-scale volcanic event involving the eruption of vast amounts of basaltic lava that covers extensive areas, such as the Deccan Traps.

Geothermal Energy—Heat energy generated and stored in the Earth's interior that can be harnessed for electricity production or heating.

Hot Spot—A localized region in the mantle where hot material rises, often producing volcanic activity at the surface even far from plate boundaries.

Hydrothermal Vent—A fissure on the seafloor that releases geothermally heated, mineral-rich water, creating habitats for specialized ecosystems.

Igneous—Relating to rocks formed through the cooling and solidification of magma or lava.

Ignimbrite—A rock formed by the deposition and welding together of volcanic ash, pumice, and other debris from a pyroclastic flow.

In Situ—Describes geological materials or fossils that are found in the original location where they were deposited.

Intrusion—The process by which magma penetrates and solidifies within preexisting rock layers, forming intrusive igneous bodies.

Intrusive Igneous Rock—Rock that forms from magma cooling slowly beneath the Earth's surface, typically resulting in coarse-grained textures (e.g., granite).

Island Arc—A curved chain of volcanic islands that forms above a subduction zone at a convergent plate boundary.

Isostatic Rebound—The gradual uplift of Earth's crust following the removal of a heavy load (such as glacial ice) as the mantle readjusts.

Jim Hutton—Refers to James Hutton, an 18th-century Scottish geologist known as the "father of modern geology" for his contributions to understanding Earth's deep-time processes.

Kīpuka—An isolated patch of older land or vegetation surrounded by newer lava flows, often forming islands within volcanic fields.

Laccolith—A dome-shaped intrusive igneous body formed when magma is injected between layers of sedimentary rock, causing the overlying strata to arch upward.

Lahar—A destructive volcanic mudflow composed of a slurry of ash, rock fragments, and water, frequently triggered by heavy rainfall or rapid melting of snow/ice.

Lava—Molten rock expelled from a volcano during an eruption; once cooled, it solidifies to form extrusive igneous rock.

Lava Dome—A rounded, mound-like volcanic feature created by the slow extrusion of highly viscous lava that piles up around the vent.

Lava Tube—A natural tunnel formed by flowing lava beneath the surface; it remains after the exterior cools and solidifies, preserving the shape of the lava channel.

Lithosphere—The rigid outer layer of the Earth, consisting of the crust and the uppermost portion of the mantle.

Mafic—Describing igneous rocks or magma rich in magnesium and iron, typically dark in color and lower in silica than felsic rocks.

Magma—Molten rock stored beneath the Earth's surface; its composition and cooling history determine the type of igneous rock that forms.

Magma Chamber—A reservoir of molten rock beneath a volcano where magma accumulates prior to eruption.

Mantle Plume—An upwelling of abnormally hot rock from deep within the mantle that can lead to surface volcanic activity, often associated with hot spots.

Mid-Ocean Ridge Basalt (MORB)—Basalt that forms at mid-ocean ridges as tectonic plates diverge, allowing magma to rise and cool to form new oceanic crust.

Moberg Ridge—A volcanic formation produced by a subglacial eruption that creates both pillow lavas and fragmented volcanic rock before breaching the overlying ice.

Mohorovičić Discontinuity Line (Moho)—The boundary between the Earth's crust and the underlying mantle, marked by a distinct change in seismic wave velocities.

Obsidian—A naturally occurring volcanic glass formed when felsic lava cools so rapidly that crystals do not have time to form, resulting in a smooth, glassy texture.

Ocean Island Basalt (OIB)—Basalt erupted from oceanic islands typically associated with mantle hotspots, often differing in composition from mid-ocean ridge basalt.

Pahoehoe—A smooth, ropy type of basaltic lava flow typically formed by low-viscosity lava, contrasting with the rough a'a flows.

Pangea—A supercontinent that existed during the late Paleozoic and early Mesozoic eras, which eventually fragmented to form today's continents.

Petrography—The study and classification of rocks, primarily using microscopic examination of thin sections to reveal mineral composition and texture.

Pillow Lava—Basaltic lava that cools rapidly on contact with water, forming rounded, pillow-shaped structures.

Plate Tectonics—The scientific theory that describes the movement of Earth's lithospheric plates and explains many geological phenomena, including earthquakes, volcanoes, and mountain formation.

Pluton—A large body of intrusive igneous rock that crystallized slowly from magma deep within the Earth's crust.

Protoplanet—A planetary embryo formed during the early stages of planet formation through the accretion of dust and smaller bodies in a protoplanetary disk.

Pumice—A highly vesicular, light, and porous volcanic rock produced during explosive eruptions; it often floats on water due to its low density.

Pyroclastic Flow—A fast-moving current of hot gas and volcanic fragments (ash, pumice, and rock) that rushes down the slopes of a volcano during an explosive eruption.

Rhyolite—An extrusive igneous rock with a high silica content, typically light in color and compositionally similar to granite.

Rootless Cone—A volcanic cone formed by the explosive interaction of lava with wet ground or water, which is not directly connected to a magma chamber.

Seamount—An underwater mountain, usually of volcanic origin, that rises from the ocean floor but does not break the surface.

Seismic Tomography—A technique that uses seismic waves to create three-dimensional images of Earth's interior, helping to reveal variations in composition and temperature.

Seismicity—The frequency, intensity, and distribution of earthquakes in a given region over time.

Shield Volcano—A broad, gently sloping volcano built primarily by the eruption of low-viscosity basaltic lava flows.

Sill—A tabular intrusive igneous body that forms when magma intrudes parallel to existing rock layers (bedding planes) and solidifies between them.

Stratovolcano—Also known as a composite volcano, it is constructed from alternating layers of lava flows, ash, and volcanic debris, formed by both explosive and effusive eruptions.

Subduction Zone—A convergent plate boundary where one tectonic plate is forced beneath another into the mantle, often associated with earthquakes and volcanic activity.

Supervolcano—A volcano that is capable of producing an eruption thousands of times larger than a typical volcanic eruption, ejecting more than 1,000 cubic kilometers of material.

Tephrochronology—The dating and correlation of discrete layers of volcanic ash (tephra) within sedimentary sequences to reconstruct the timing of past volcanic events.

Tsunami—A series of large ocean waves generated by underwater disturbances—such as earthquakes, volcanic eruptions, or landslides—that can cause widespread coastal damage.

Tuya—A flat-topped, steep-sided volcanic formation created when lava erupts beneath a glacier or ice sheet, cooling rapidly and forming a distinctive shape.

Uranium Disequilibrium Series—A sequence of radioactive decay processes involving uranium isotopes used in geochronology to date rocks and sediments.

Viscosity—A measure of a fluid's resistance to flow; in volcanology, it is crucial in determining how lava moves and the type of volcanic landforms that develop.

Volcanic Explosivity Index (VEI)—A scale that quantifies the explosiveness of volcanic eruptions, ranging from 0 (effusive) to 8 (super eruptions), based on eruption volume and intensity.

Volcano—An opening in Earth's surface where magma, gases, and volcanic ash are expelled during an eruption.

Xenolith—A fragment of preexisting rock incorporated into an igneous rock during the ascent of magma, providing clues about Earth's deeper composition.

APPENDIX

This selection of the "100 Most Active Volcanoes in the World" is based on recent activity, historical eruptions, and potential threat levels. Highlighted in bold on this list are 16 volcanoes identified by the International Association of Volcanology and Chemistry of the Earth's Interior (IAVCEI) as posing a significant threat due to their history of large, destructive eruptions and proximity to populated areas. These 16 volcanoes are sometimes referred to as the "Decade Volcanoes." Those volcano names on this list that are followed by an asterisk indicate the active volcanoes that pose the most threat in North America, according to the United States Geological Society (USGS). The list is organized by continent and common volcano name in alphabetical order, with the country (or state) of origin.

VOLCANOES OF NORTH AMERICA

Colima Volcano, Mexico. Colima Volcano (3,280 m), located in western Mexico, is part of a volcanic complex that results from the subduction of the Cocos Plate beneath the North American plate. This tectonic setting lends itself to explosive eruption styles, resulting in the eruption of Colima Volcano over 40 times since 1576, making it one of the most active volcanoes in Mexico. Notable eruptions occurred in 1913, 1940, and 1991, each contributing to the volcano's cone shape, altering the surrounding landscape, and blanketing nearby cities Colima City and Manzanillo.

Crater Lake, Oregon, USA. Crater Lake, located in southern Oregon, is the result of the collapse of Mount Mazama. This stratovolcano erupted approximately 7,700 years ago—one of the most powerful eruptions to occur in North America. Over time, the caldera filled with rainwater and snowmelt, forming Crater Lake, known for its stunning blue color and clarity. The lake is approximately 592 m deep, making it the deepest in the United States. Subsequent volcanic activity has included the formation of several islands, such as Wizard Island, which features a cinder cone.

Glacier Peak, Washington, USA. Glacier Peak, located in Washington State, is one of the five active volcanoes in the Cascade Mountain Range. Its most significant eruptions occurred approximately 13,000 and 6,000 years ago, producing extensive pyroclastic flows and ash deposits that have been traced as far as the midwestern United States. The 6,000-year-old eruption is particularly notable for its scale, as it created a large volcanic cone and significantly impacted local ecosystems. Subsequent more minor eruptions were recorded in the late 19th century, with the last significant activity occurring in 1956, when a series of explosive eruptions sent ash plumes into the atmosphere. Glacier Peak is recognized for its glacial cover, which can influence its volcanic activity and increase the risk of lahars during eruptions.

Kilauea, Hawaii, USA*. Kilauea, located on the Big Island of Hawai'i, is one of the most active volcanoes in the world, with a volcanic history spanning over 300,000 years. The long-term volcanic activity at Kilauea is supported by the Hawaiian Hot Spot that sits underneath the Pacific tectonic plate. It has erupted continuously since 1983. The most notable recent eruption occurred in 2018, when a series of explosive events and massive lava flows resulted in the destruction of homes and the expansion of the island by approximately 1,500 acres

(607 hectares). Kilauea's summit caldera, about 3 km in diameter, hosts a lava lake, which provides valuable insights into volcanic processes. Kilauea is characterized by its relatively gentle effusive eruptions, primarily producing basaltic lava.

Lassen Peak, California, USA*. Lassen Peak is a noteworthy stratovolcano in Northern California that is part of the Cascade Mountain Range and stands at 3,187 m. Formed over approximately 650,000 years, it is best known for its eruptive phase from 1914 to 1921, which included explosive eruptions and the growth of a lava dome. These events produced significant flows of andesitic lava and reshaped the surrounding landscape. Although there have been no significant eruptions since 1921, Lassen Peak is closely monitored by the USGS for signs of volcanic unrest, highlighting its potential for future activity.

Long Valley Caldera, California, USA*. Located in Eastern California, Long Valley is a massive volcanic caldera that formed from a colossal explosive eruption approximately 760,000 years ago. This eruption was one of the largest in North America, ejecting an estimated 140 km^3 of material and leading to the subsidence of the land, which created the caldera that now spans about 32 km long and 16 km wide. The eruption produced the Bishop Tuff, a significant volcanic rock formation that is crucial for understanding the region's geological history.

Makushin Volcano, Alaska, USA*. Makushin is an ice-covered stratovolcano located in the Aleutian Volcanic Arc. Makushin is the highest summit in the Aleutian Islands at 2,036 m. This volcano formed in the early Pleistocene with extensive flows composed of andesitic and basaltic lava that enlarged the island by several kilometers along the northwest coast. Part of Alaska's 52 historically active volcanoes, Makushin has erupted numerous times over the past thousands of years, with its most recent eruption occurring in 1995. The January 1995 eruption produced an ash cloud that rose roughly 2.5 km above the volcano.

Mammoth Mountain, California, USA. Mammoth Mountain (3,369 m), located in California's Sierra Nevada, is a prominent volcanic structure created over 200,000 years ago. The mountain is primarily composed of lava flows and ash deposits from 25 unique eruptions, with the most significant activity occurring between 200,000 and 50,000 years ago. The last known eruption took place approximately 300 to 400 years ago, producing the prominent volcanic features visible today, including the famous volcanic cones and craters. Mammoth Mountain is characterized by its extensive geothermal activity, evidenced by hot springs and fumaroles, which indicate the presence of magma beneath the surface.

Mauna Loa, Hawaii, USA. Mauna Loa, located on the Big Island of Hawaii, is the largest active volcano in the world, covering an area of approximately 5,271 km^2. Its volcanic history spans over 700,000 years, with the most recent eruption occurring from November 27 to December 10, 2022. This eruption produced lava flows that traveled downslope but did not threaten populated areas. Mauna Loa is characterized by its shield volcano structure, formed primarily by low-viscosity basaltic lava that results in gentle slopes. Historically, the volcano has erupted approximately 33 times since its first well-documented eruption in 1843. The eruptions have varied in intensity, with some producing extensive lava flows that significantly altered the landscape. The summit caldera, Moku'āweoweo, measures about 5 km in length.

Mount Baker, Washington, USA*. Mount Baker is located in Washington and is one of the youngest volcanoes in the Cascade Mountain Range. The stratovolcano formed during the Last Ice Age, roughly 15,000 years ago, and has been steadily growing ever since. Its most significant eruption occurred in 1843, which produced a massive ash plume and pyroclastic flows, marking the first recorded eruption in the area. Subsequent eruptions in the late 19th and early 20th centuries, notably in 1880 and 1921, further reshaped the landscape, contributing to the formation of new lava domes and significant ash deposits.

The volcano remains active, with minor eruptive episodes and seismic activity reported intermittently.

Mount Churchill, Alaska, USA. Mount Churchill is located on Umnak Island in the Aleutian Range of Alaska. The most notable event in its eruptive history occurred in 1985, when the volcano experienced a significant eruption that began in late May and continued intermittently until early June. During this eruption, explosive activity generated ash plumes that reached heights of approximately 4,600 m, dispersing ash over a wide area, including parts of the Aleutian Islands. The eruption produced both ash fall and volcanic gases, creating hazards that affected air quality and posed risks to aviation. Although Mount Churchill has a relatively sparse eruptive history overall, with limited documentation of earlier eruptions, the 1985 event emphasized its potential for explosive activity.

Mount Cleveland, Alaska, USA. This stratovolcano is located in Alaska, in the central part of the Aleutian Island Arc, and makes up half of the island it sits on, called Chuginadak. Mount Cleveland has erupted 21 times over the past 250 years. The most recent eruption at Mount Cleveland occurred in 2011; however, the volcano had erupted three times in 2009 and twice in 2010 prior to that. Mount Cleveland is infamous for its large ash plumes—every eruption since 2000 has affected air traffic. Although the exact date that Mount Cleveland was formed is not specific, it can be estimated to be around 10,000 years ago, indicating it is a Holocene volcano. The earliest records of this volcano were from the Aleut people, who named it Chuginadak after the Aleut fire goddess.

Mount Hood, Oregon, USA. The volcanic activity at Mount Hood (Oregon) began 500,000 years ago. Its most recent significant eruption occurred around 1907, producing a series of small eruptions and ash plumes that impacted the surrounding areas. Prior to this, Mount Hood experienced notable activity approximately 1,500 years ago, resulting in the formation of lava flows and volcanic debris that can still be observed today. The volcano's summit reaches an elevation of

3,426 meters, making it the highest peak in Oregon. Its proximity to urban areas, including Portland, raises concerns about future eruptions and associated risks, prompting continuous monitoring by the USGS.

Mount Meager, Canada. Mount Meager is located in Canada in the Pacific Range of the coastal mountains in British Columbia. Mount Meager Massif is a group of six volcanic peaks. Mount Meager itself is the second-highest peak at 2,690 m. Mount Meager Massif is part of the Cascade Volcanic Arc in western North America. Although its last eruption was 2,000 years ago, the Massif could still produce a devastating eruption, as it still holds the title for the most enormous explosion in Canadian volcanic history with its eruption 2,400 years ago. During that eruption, the Massif was formed. A 2010 landslide created massive debris that devastated the surrounding area.

Mount Okmok, Alaska, USA. Mount Okmok, located in the Aleutian Islands of Alaska, is a stratovolcano with a complex volcanic history marked by significant eruptions. The most notable event occurred in 2008, when a series of eruptions produced ash plumes that disrupted air travel and led to the evacuation of nearby communities. This eruption created a new lava dome within the caldera, which had previously formed during a massive eruption approximately 2,000 years ago, believed to be one of the largest in the region. The caldera itself measures about 10 km in diameter, indicating the scale of past explosive activity. Prior to the 2008 eruption, Mount Okmok had been relatively quiet for several decades.

Mount Pavlof, Alaska, USA. Mount Pavlof (2,508 m tall) is located on the Aleutian Peninsula of Alaska. As a stratovolcano, it features a conical shape formed by layers of lava flows and volcanic debris resulting from the subduction of the Pacific Plate beneath the North American Plate, forming the Aleutian Volcanic Arc. Mount Pavlof is the most active volcano in the United States since 1980, with eruptions recorded in 1980, 1981, 1983, 1986–1988, 1996–1997, 2007, 2013, 2014,

2016, 2021–2022, and 2024. Anticipated eruptions of Mount Pavlof are considered a high threat to air travel because its lava is silica-rich, containing greater than 53% silica (SiO_2).

Mount Rainier, Washington, USA*. Mount Rainer is a glaciated stratovolcano in Washington and is the highest peak in the Cascade Range, standing at 4,392 m. Formed over the past 500,000 years, Mount Rainier's geological history includes numerous eruptive phases, with the last significant eruption occurring around 1894. During the last major eruption, lahars flowed down the Puyallup River, burying parts of the valley under several feet of debris. In more recent history, a lahar in 2009 was triggered by the melting of snow and ice, which led to debris flows in the river valleys surrounding the mountain, threatening nearby communities such as Orting and Puyallup. Mount Rainier is considered one of the most hazardous volcanoes in the world due to its high likelihood of eruption in the near future and its proximity to the urban city of Seattle, where nearly one million people reside.

Mount Redoubt, Alaska, USA*. Mount Redoubt is located at the northeast end of the Aleutian Volcanic Arc and is considered the most potentially dangerous volcano in Alaska. It formed around 890,000 years ago with dacite explosive eruptions. Cone-building eruptions began around 340,000 years ago, but it was not until the later cone-building stage that Mount Redoubt's magma supply increased in silica content. The 1989 eruption spewed volcanic ash to a height of 13.2 km, affecting the flight path of a Boeing 747 aircraft passing by. Additionally, the 1989 eruption impacted the Cook Inlet Region's oil production, transportation, and the local population. At the time, the Mount Redoubt 1989 eruption was the second most costly eruption in the United States at $160 million.

Mount Saint Helens, Washington, USA*. Mount Saint Helens is located in the Cascade Mountain Range in southwestern Washington. The eruption on May 18, 1980, still ranks as one of the worst volcanic eruptions in modern history. It began in March 1980 with explosive steam eruptions. On the morning of May 18, a 5.1 magnitude earthquake set off a massive landslide on the north side of the volcano. This avalanche was quickly followed by a lateral air blast that carried 8 km of superheated ash and stone. This blast alone reached temperatures of 348°C and speeds up to 480 km per hour. A simultaneous vertical eruption of gas and ash formed a column about 25 km high. Complete darkness then occurred about 400 km northeast of Mount Saint Helens. This event killed 57 people, thousands of animals, and widespread plant life.

Mount Shasta, California, USA*. Mount Shasta is a stratovolcano located in Northern California's portion of the Cascade Range that stands at an elevation of 4,422 m. It has a complex geological history characterized by a series of eruptive phases that began approximately 300,000 years ago. Mount Shasta is primarily composed of andesite, basalt, and dacite, reflecting its diverse eruptive styles. The formation of Mount Shasta can be divided into several key phases. Early eruptions produced basaltic lava flows that formed the foundation of the volcano. Subsequent cone-building eruptions, particularly around 10,000 years ago, led to the development of its iconic cone shape through the extrusion of andesitic lava. The most recent major eruption occurred in 1786, producing substantial ashfall. Since then, the volcano has shown intermittent activity, including minor steam eruptions.

Mount Spurr, Alaska, USA*. Mount Spurr is located in Alaska in the Aleutian Volcanic Arc and is the highest peak in the arc at 3,374 m. Mount Spurr is defined by its large lava dome located at the volcano's center as well as a 5 km horseshoe-shaped caldera that opens to the south side. This caldera likely formed during the late Pleistocene. The most recent eruptions occurred in 1953 and 1992, both originating from a vent approximately 3.5 km below the surface. Both of these eruptions caused ash plumes that affected nearby cities and, with the 1992 eruption, air traffic.

Novarupta, Alaska, USA. Novarupta, located on the Alaska Peninsula, is renowned for the cataclysmic eruption in June

1912, which ranks among the most significant eruptions of the 20th century. This eruption produced approximately 30 times the ash volume of the infamous 1980 Mount Saint Helens eruption, blanketing the surrounding areas in ash. The event created a new caldera, with the eruption lasting about 60 hours and resulting in the formation of the Novarupta Lava Dome. Prior to 1912, Novarupta had exhibited sporadic activity; however, the eruption marked a pivotal moment in its volcanic history. Since then, the volcano has remained relatively quiet, with only minor eruptions reported in the following decades.

Paricutin Volcano, Mexico. Paricutin Volcano, located in the Mexican state of Michoacán, is renowned for its unique volcanic history, particularly because it emerged in a farmer's field in 1943, marking one of the first volcanic eruptions observed in real time. The eruption began with a series of explosive events that produced ash plumes and lava fountains, ultimately resulting in the formation of a cinder cone that reached an elevation of approximately 424 m. Over the next 9 years, Paricutin erupted continuously, with lava flows burying two nearby towns, San Juan Parangaricutiro and Paricutin. The eruption concluded in 1952.

Popocatépetl, Mexico. Popocatépetl, located in central Mexico, is one of the country's tallest peaks (5,426 m) and most active volcanoes. Its geological history spans over 730,000 years and encompasses various eruptive phases characterized by explosive Plinian eruptions (1195 BC, 800 BC, AD 823), which produced predominantly andesitic lava flows. More recent notable eruptions include those in 1947 and 1994, which produced substantial ashfall and pyroclastic flows affecting nearby communities. The 2013 eruption, which launched ash to heights exceeding 5 km, prompted evacuations and significant monitoring by the Mexican government due to the potential hazards posed to surrounding towns. Popocatépetl's ongoing activity, including the formation of a lava dome within its summit crater, underscores the importance of continuous geological monitoring to ensure the safety of the millions of people living in its vicinity. Its striking appearance and cultural significance make Popocatépetl not only a geological landmark but also a vital part of Mexico's heritage.

Three Sisters Volcanoes, Oregon, USA*. The Three Sisters Volcanoes, located in the Cascade Range of Oregon, comprise South Sister, Middle Sister, and North Sister, each with a distinct volcanic history. South Sister, the youngest, last erupted around 2,000 years ago, creating a significant lava flow and ash deposits. Middle Sister has been relatively quiet, with its last known eruption occurring approximately 1,000 years ago. In contrast, North Sister is considered dormant, with no eruptions recorded in the last few thousand years, but it features prominent glacial erosion that reveals its volcanic past. Collectively, the Three Sisters have produced a range of volcanic materials, including andesitic and rhyolitic lavas, pyroclastic deposits, and significant glacial activity, which continues to shape the landscape. The region is closely monitored due to its potential for future eruptions and its proximity to populated areas. The Three Sisters are not only geologically significant but also a popular destination for recreation and research in volcanic processes.

Yellowstone Supervolcano, Wyoming, USA. The Yellowstone Supervolcano, located primarily in Wyoming, is one of the largest active volcanic systems in the world (i.e., Yellowstone Plateau Volcanic Field, covering 17,000 km^2) with a volcanic history spanning over 2.1 million years. It has experienced three major eruptions, the most significant occurring approximately 640,000 years ago, which created the Yellowstone Caldera, measuring about 70 km in diameter. This eruption expelled an estimated 1,000 km^3 of material, profoundly altering the landscape and climate. Subsequent eruptions occurred around 1.3 million and 2.1 million years ago, each contributing to the caldera's formation. The area is characterized by geothermal features, including geysers, hot springs, and fumaroles, which indicate ongoing volcanic activity. The most famous geyser, Old Faithful, erupts regularly, showcasing the underlying geothermal dynamics.

VOLCANOES OF CENTRAL AMERICA

Arenal Volcano, Costa Rica. Volcano Arenal, located in northwestern Costa Rica, was initially formed 7,500 years ago and is known for its picturesque conical appearance that rises 1,670 m. After 400 years of dormancy, Arenal Volcano experienced a significant eruption in 1968 that continued for the next 42 years. The initial eruption not only was significant for its explosive power—producing lava flows, ash clouds, and pyroclastic flows—but also destroyed several nearby towns, including the village of Tabacón. A tourist town called La Fortuna was created on the eastern side of Arenal to view ongoing lava fountains that captivated locals and visitors. While Arenal has shown signs of decreased activity, the area continues to enjoy the benefits of geothermal pools and hot springs for recreation.

Concepción Volcano, Nicaragua. Concepción is one of two volcanoes (the other being Maderas) located on Isla de Ometepe within Lake Nicaragua in Nicaragua. Concepción Volcano is a stratovolcano that stands at 1,610 meters. Its geological history spans over 50,000 years and is characterized by both explosive and effusive volcanic activity. Concepción Volcano's most recent significant eruptions occurred in the 20th century, with notable activity in 1957 and 2005.

Fuego Volcano, Guatemala. Fuego Volcano, considered by many to be the most visually stunning volcano in Guatemala, towers at an impressive 3,763 m and is a prominent feature of the Central American landscape. The geological history of Fuego is marked by a series of explosive eruptions, with activity recorded as far back as the 16th century. Notably, the volcano has erupted over 60 times since 1542, showcasing its volatility. One of the most significant eruptions occurred in June 2018, when a catastrophic explosion generated pyroclastic flows that devastated the nearby town of San Miguel Los Lotes, resulting in tragic loss of life and widespread destruction. The eruption sent ash plumes soaring 9,144 m into the atmosphere, affecting air travel and blanketing surrounding areas in volcanic ash.

Irazú, Costa Rica. Irazú Volcano, located in the central highlands of Costa Rica, is the highest active volcano in the country, with a summit elevation of 3,432 m. Its volcanic history spans over 1,000 years, with 23 eruptions documented since the 18th century, ranking VEI 1 to 3. The most notable eruption occurred between 1963 and 1965, when a series of explosive events produced ash plumes that reached altitudes up to 10 km, affecting air quality and disrupting daily life in San Jose. This eruption created a new crater that hosts a lake today.

Masaya Volcano, Nicaragua. Masaya Volcano is one of the most active volcanoes in Central America, dating back approximately 20,000 years. As a prominent stratovolcano, the most significant activity began in the 16th century, with eruptions recorded as early as 1520. Today, the volcano is characterized by its continuous activity, with eruptions typically occurring every few years. Noteworthy eruptions in the 20th century include those in 1977 and 2008, which produced lava fountains and significant ash emissions, affecting local air quality and visibility. Masaya's unique geological features include several active craters, such as the Santiago crater, which is often filled with a lava lake.

Momotombo Volcano, Nicaragua. Momotombo, situated near Lake Xolotlán in Nicaragua, is a stratovolcano with a geological history spanning over 10,000 years. Its most significant eruptions have occurred throughout the 20th century, with notable activity in 1905 and 2015. The 1905 eruption was particularly dramatic, resulting in a large ash cloud and extensive lava flows, marking it as one of the most significant volcanic events in Nicaraguan history. A geothermal field on the southern flank is harnessed for electricity.

Pacaya Volcano, Guatemala. Pacaya Volcano rises 2,552 m over Guatemala City. Its geological history, spanning over 23,000 years, is marked by frequent eruptions that have shaped the surrounding landscape and impacted local communities. A dramatic eruption in 1965 produced a lava flow that traveled down the southwest flank, threatening nearby towns. More recently, in 2010, a series of explosive eruptions sent ash plumes high into the atmosphere, affecting air travel and blanketing the region in volcanic ash. The 2014 eruption also garnered attention, as it generated lava flows that advanced toward the nearby village of El Patrocinio, prompting evacuations and showcasing the volcano's dynamic nature.

Poás, Costa Rica. Poás Volcano, located in central Costa Rica, is an active stratovolcano that hosts a 1,500 m crater that contains a sulfuric acid lake (Laguna Caliente) considered to be one of the most acidic environments on the planet with a pH equal to zero. The volcano's eruptive history spans thousands of years, with significant activity recorded since the Spanish conquest in the 16th century. Notable eruptions occurred in 1953, producing ash clouds that affected nearby communities and resulted in the closure of the national park. A series of eruptions in 2014, 2017, and 2025 included explosive events that generated ash plumes and volcanic gas emissions, prompting increased monitoring and temporary evacuations. The summit reaches an elevation of 2,708 m and is characterized by its lush vegetation and diverse ecosystems.

San Cristóbal Volcano, Nicaragua. San Cristóbal, the highest volcano in Nicaragua at 1,745 m, is a prominent stratovolcano located in the northwestern part of the country. Its geological history spans over 30,000 years, marked by a combination of explosive eruptions and lava flows that have shaped the surrounding landscape. The first recorded eruption occurred in 1527, with significant activity noted in the 20th and 21st centuries, including eruptions in 2006 and 2015, which produced ash plumes that affected air quality and local communities.

Santa María Volcano, Guatemala. Santa Maria, rising 3,772 meters in western Guatemala, is a stratovolcano with a geological history spanning over 30,000 years. Notable for its explosive activity, the volcano's most powerful eruption occurred in 1902, producing an ash plume that soared 20 km into the atmosphere, leading to widespread ashfall and agricultural disruption. This eruption also generated the Santiaguito lava dome, which remains an active volcanic site, exhibiting frequent eruptions that impact nearby communities, such as Quetzaltenango.

VOLCANOES OF SOUTH AMERICA

Alcedo Volcano, Ecuador. Alcedo Volcano, located in the center of Isabela Island, is one of six coalescing shield volcanoes that make up the Galapagos Island chain. The Alcedo Volcano is the result of the Galapagos Hot Spot and formed about 313,000 years ago. In terms of uniqueness, Alcedo is the only volcano within the chain to have erupted with both rhyolite and basalt. Alcedo's last eruption was in 1993.

Calbuco Volcano, Chile. Calbuco is located in southern Chile in the Los Lagos Region. Calbuco is characterized as a stratovolcano with andesitic lava flows, with its summit reaching approximately 400 m in height. Calbuco had roughly 36 eruptions during the Holocene. The most notable eruption was the 1893–1895 eruption. This was the largest eruption recorded in southern Chile, with debris ejected up to 8 km and massive lahars. This volcano erupted in April 2015 and ranked a VEI 4. This was Calbuco's first eruption since 1972.

Cerro Negro Volcano, Nicaragua. Cerro Negro is located in Nicaragua in the Cordillera de los Maribios Mountain range. The youngest volcano in Central America, Cerro Negro, was first formed in April 1850 as a Strombolian eruption, ranking a VEI 2. This volcano has erupted 13 times since 1850, with the last event in 1999.

Copahue Volcano, Chile and Argentina. Copahue is located on the border of Chile and Argentina. A distinct feature of Copahue is the nine volcanic craters along a 2 km fissure found at its summit. The easternmost crater is historically the most active and continues to produce fumarolic activity. The modern volcano caldera formed roughly 0.6 million years ago and became active roughly 1.2 million years ago. The most recent eruptions, in 2020 and 2021, saw SO_2 emissions of 4,684 tons/day.

Cotopaxi Volcano, Ecuador. Cotopaxi is located in Ecuador and is situated in the Cordillera Central of the Andes mountains. A violent eruption occurred in 1877, with lahars flowing downslope at 100 km/h that destroyed the town of Latacunga. The most recent eruption was in November 2022, producing an ash plume that reached 1.8 km above its crater.

Galeras Volcano, Columbia. Galeras, located in southwestern Columbia near the city of Pasto, is an active stratovolcano. Its first eruptions date back over 12,000 years, and the volcano has undergone numerous cycles of explosive and effusive activity, shaping the surrounding landscape. The recorded history of Galeras includes significant eruptions, particularly in 1993 and 2010, generating pyroclastic flows and ash fallout that posed serious risks to nearby populations.

Mitsi Volcano, Peru. Mitsi is located in southern Peru and is situated in the Andes Mountains. This stratovolcano formed 112,000 years ago; however, Mitsi has erupted 12 times within the past 50,000 years, forming a caldera at 13,700 and a scoria cone at 11,000 years ago. An eruption that occurred approximately 2,000 years ago is considered the worst eruption ever to have taken place in Peru, with an estimated Volcanic Explosivity Index (VEI) of 5.

Nevado del Huila Volcano, Columbia. Nevado Del Huila is an andesitic volcano formed on top of the Ibagué Batholith. The eruptive history of Nevado del Huila began 1.5 million years ago and is divided into three stages: the Pre-Huila Stage (ending approximately 100,000 years ago), the Ancient Huila Stage (ending during the last glaciation), and the Recent Huila Stage (beginning around 11,000–10,000 years ago). The most notable eruption was in April 2007, when it erupted twice, causing avalanches. During a 2008 eruption, authorities quickly evacuated nearby towns to save 12,000 lives from a fast-moving lahar.

Nevado del Ruíz Volcano, Columbia. Nevado del Ruíz Volcano is a stratovolcano that is part of the Ruiz-Tolima volcanic mastiff. The mastiff itself sits on multiple faults, some of which are still active. An eruption in 1985 triggered a massive lahar that buried the entire town of Amerero, killing 23,000 people.

Ojos del Salado, Chile and Argentinian. Ojos del Salado is located on the border of Chile and Argentina in the Andes Mountains. Ojos Del Salado is both the world's highest volcano and crater lake, at 6,480 m, and the highest peak in Chile. The upper divisions of the volcano consist of several overlapping lava domes, lava flows, and volcanic craters. Although there is no record of the eruptions at Ojos del Salado, its current fumarolic activity suggests that it is not dormant.

Osorno Volcano, Chile. Osorno is located in southern Chile in the Los Lagos region. The Natives had several names for the volcano, such as Purailla, Hueñauca, and Guanaca, among others, before the settlers in this region gave it a name. Osorno sits within La Picada, a 6-km-wide caldera that was formed 250,000 years earlier. Osorno is one of the most active volcanoes in the southern Andes mountains, with 11 eruptions recorded between 1575 and 1869. An eruption on January 19, 1835, was witnessed and recorded by Charles Darwin during one of his voyages on the HMS Beagle.

Puyehue-Cordón Caulle, Chile. Located in the Chilean Andes, Puyehue-Cordón Caulle is a pair of coalesced volcanic edifices that form a 300,000-year-old mountain massif. The most notable eruption was in 1960, following the largest earthquake recorded in history, known as the Great Chilean Earthquake, with a magnitude of 9.5. Geothermal springs provide evidence of Puyehue-Cordón Caulle activity and are part of the largest field in the southern Andes Volcanic Zone at 36 km by 13 km. The most recent eruption was in 2010–2011.

Reventador Volcano, Ecuador. Reventador is a 400,000-year-old stratovolcano located in Ecuador in the eastern Andes Mountains. Reventador features a 4-km-wide andesitic caldera that formed approximately 30,000 years ago. Since 1541, the volcano has erupted 25 times, the worst being in 2002 with an ash plume over 17 km tall.

Sangay Volcano, Ecuador. Sangay Volcano is located in Central Ecuador and lies on the eastern edge of the Andes Mountains. This volcano has been highly active since 1628 and is notable for producing Strombolian-style eruptions with lava fountains over 100 m tall, as was reported in 2020.

Ubinas Volcano, Peru. Ubinas, a stratovolcano 5,672 m tall, is located in southern Peru and is part of the central volcanic region of the Andes Mountains. Eruptions that occurred from 2015 to 2017 produced large amounts of sulfur dioxide, with daily averages of up to 3,700 tons per day.

Villarrica Volcano, Chile. Villarrica Volcano is located in central Chile, about 750 km from the capital city of Santiago. The stratovolcano is one of the most active volcanoes in Chile. Two calderas have formed at Villarrica, the oldest being 10,000 years old and the youngest 3,500 years old. Villarrica currently has an active lava lake and is one of nine volcanoes monitored for carbon dioxide and sulfur dioxide emissions by the Deep Earth Carbon Degassing project. Eruptions in 1964 and 1971 produced massive lahars.

VOLCANOES OF EUROPE

Askja, Iceland. Askja is a stratovolcano located in the central highlands of Iceland. Formed during a series of explosive eruptions, a caldera that measures about 50 km in diameter was created approximately 10,000 years ago. An eruption in 1875 produced substantial ash fallout that affected agriculture in Iceland and even reached as far as Norway. This eruption led to the creation of the Öskjuvatn Lake within the caldera, which is now one of the largest lakes in Iceland. Another significant event took place in 1961–1962, when Askja erupted explosively, resulting in a new lava flow and ash emissions that impacted the surrounding environment. The most recent eruptive activity occurred in 2014, marked by a series of small eruptions and increased geothermal activity, prompting close monitoring by Icelandic authorities.

Bárđarbunga, Iceland. Bárđarbunga is a stratovolcano located beneath the Vatnajökull glacier in central Iceland. With a peak elevation of 2,009 m, Bárđarbunga is one of Iceland's tallest volcanoes. The most significant eruption in recent history occurred from August 2014 to February 2015. This eruption was part of a rifting event that released large volumes of lava, creating the Holuhraun lava field, which covered approximately 85 km^2. The eruption produced impressive lava fountains and emitted substantial amounts of volcanic gas, impacting air quality and aviation in the region. Prior to the 2014–2015 eruption, Bardarbunga experienced a series of eruptions in the 20th century.

Campi Flegrei, Italy. Campi Flegrei, also referred to as the Phlegraean Fields, is a large volcanic caldera located near Naples, Italy. Formed approximately 39,000 years ago, it is the result of a series of explosive eruptions that shaped its landscape, characterized by craters, fumaroles, and hot springs. One of the most significant eruptions occurred around 15,000 years ago. Known as the Campanian Ignimbrite eruption, it produced extensive ash deposits and contributed to the formation of the caldera itself. The eruption was one of the largest in Europe, affecting climate and human populations across the region. Historically, the last major eruption occurred in 1538, resulting in the formation of the Monte Nuovo cone, which emerged within a matter of days. This eruption destroyed several villages and displaced thousands of people.

Eyjafjallajökull, Iceland. Eyjafjallajökull is a prominent stratovolcano located in southern Iceland that became famous for its 2010 eruption, which gained international attention due to its massive ash plume, which disrupted air travel across Europe for several weeks. The 2010 eruption produced an estimated 250 megatons of ash, which spread across the continent and led to the grounding of over 100,000 flights. Previously, Eyjafjallajökull had minor eruptions in 1821–1823.

Hekla, Iceland. Hekla is located in the south-central region of the country. With a geological history that dates back about 11,000 years, this stratovolcano is often referred to as the "Gateway to Hell" in medieval literature, reflecting its reputation for explosive activity. In recent history, Hekla is known for having a recurrence interval of approximately 10 years, with eruptions occurring in 1948, 1960, 1970, 1980, 1991, and 2000.

Katla, Iceland. Katla is a large stratovolcano located beneath the Mýrdalsjökull glacier in southern Iceland. Its geological history spans thousands of years, with eruptions occurring approximately every 40 to 80 years, making it one of Iceland's most active volcanoes. An explosive eruption occurred in 1918 and ranked 5 on the VEI. However, the eruption of 1755 was more devastating, as it generated a mega-jökulhlaup estimated to have a discharge equivalent to that of the Amazon River, or 209,000 m^3/s.

Kraftla, Iceland. Krafla is a volcanic system located in northeastern Iceland, characterized by its geothermal fields. The area is part of the larger East Iceland Volcanic Zone and experienced nine fissure eruption phases, commonly referred to as the Kraftla Fires, from 1975 to 1984. The eruptions released large volumes of lava and produced a notable fissure eruption in 1984, resulting in the creation of the Ljósavatn lava field. Additionally, this volcanic system is linked to the Myvtan Fires, which altered the basin shape of Lake Myvtan over the course of 5 years, from 1724 to 1729. Kraftla is the site of Iceland's first geothermal energy plant that is still operational today.

Mount Etna, Italy. Mount Etna, located on the east coast of Sicily, is one of the world's most active volcanoes and the highest in Europe, reaching about 3,329 m. Its geological history dates back approximately 500,000 years and is characterized by a complex interplay of stratovolcano and shield volcano formation. Over time, it has developed a diverse landscape marked by craters, lava flows, and ash deposits. Notable eruptions include the major explosive event in 1669, which devastated the nearby town of Catania, and the 1928 eruption, which buried several villages. More recently, a fissure eruption occurred in 2025, producing a 3-km-tall ash cloud and a pyroclastic flow.

Santorini, Greece. Santorini, part of the Cyclades archipelago in Greece, is a volcanic island famous for its stunning caldera and dramatic cliffs. Its geological history is marked by a series of significant eruptions over the past 200,000 years, with the most notable being the Minoan eruption around 1600 BC. This catastrophic event is one of the most significant volcanic eruptions in recorded history, producing massive ash fallout and pyroclastic flows that devastated the Minoan civilization on the nearby island of Crete. The Minoan eruption led to the partial collapse of the island, forming the caldera seen today. Following this eruption, Santorini experienced subsequent volcanic activity, including the formation of the islets of Nea Kameni and Palea Kameni in the 20th century, which resulted from more minor eruptions in 1925, 1939, and more recently in 2006–2007. These eruptions primarily produced lava flows and minor explosive activity, contributing to the island's ongoing geological evolution.

Stromboli, Italy. Stromboli is a volcanic island in the Tyrrhenian Sea, part of Italy's Aeolian Islands, and is renowned for its persistent volcanic activity. Often referred to as the "Lighthouse of the Mediterranean," this stratovolcano has a complex geological history characterized by frequent eruptions over the past 2,000 years, primarily of basaltic lava. Notable eruptions include the 1930 explosion, which caused significant damage to the village of Stromboli, and the more recent continuous eruptive phase that began in 1932 and continues to this day. In July 2019, a significant eruption sent ash and debris into the air, prompting evacuations.

Tagoro, Spain. Tagoro is an underwater volcano located off the coast of El Hierro, one of the Canary Islands, in Spain. It lies approximately 2 km from the island and is part of a volcanic system that has been active for millions of years. The geological history of Tagoro is marked by significant eruptive activity from 2011 to 2012, which began with seismic activity that indicated magma movement beneath the ocean floor. This culminated in a submarine eruption that produced a substantial ash cloud and lava flows and resulted in the formation of a new volcanic cone.

Teide, Spain. Teide is a stratovolcano located on the island of Tenerife in the Canary Islands, Spain, and is the highest peak in Spain at 3,718 m. Its geological history spans approximately 170,000 years and is characterized by a series of eruptive phases that have shaped its impressive landscape. The largest eruption in recent times occurred in 1909, when a fissure eruption at San Juan produced significant lava flows and ash clouds, affecting local agriculture and prompting evacuations. The formation of a new volcanic cone, Montaña Chahorra, marked this eruption. Prior to this, the 1798 eruption also had a considerable impact, resulting in the basaltic lava flows that reached the sea. This volcano is a UNESCO World Heritage site.

Vesuvius, Italy. Mount Vesuvius, located near Naples, Italy, is one of the most famous and dangerous volcanoes in the world, primarily due to its explosive eruptions and proximity to densely populated areas. Vesuvius is classified as a stratovolcano, with its most significant eruption occurring in AD 79, when Vesuvius erupted catastrophically, burying the Roman cities of Pompeii and Herculaneum under volcanic ash and pumice, preserving them for centuries. This eruption released a lethal cloud of ash 33 km tall, leading to the deaths of thousands. Other significant eruptions include the 1631 and 1906 eruptions.

VOLCANOES OF AFRICA

Aluto, Ethiopia. Aluto is a volcanic complex located in the Ethiopian Rift Valley, about 200 km south of Addis Ababa, the capital of Ethiopia. This stratovolcano has a geological history characterized by a series of eruptions that occurred over the past few hundred thousand years, contributing to the formation of its distinctive landscape, which includes craters, lava lakes, and geothermal features. The most recent eruption of Aluto occurred in 1966.

Dabbahu, Ethiopia. Dabbahu, also known as the Dabbahu Volcanic Complex, is an active stratovolcano located in the northeastern part of Ethiopia within the East African Rift system. The basaltic fissure eruption of Dabbahu in 2005 was part of a larger rifting episode in the region, which caused the ground to open and resulted in the formation of a new volcanic vent.

Erta Ale, Ethiopia. Erta Ale is a stratovolcano located in the Afar Region of northeastern Ethiopia. This volcano is part of the East African Rift system. It is characterized by a lava lake that has been one of the few permanent lava lakes in the world, offering insights into basaltic volcanism. The most notable eruptions of Erta Ale occurred in 2005 and 2017. The 2005 eruption was marked by significant lava flow activity and the expansion of the lava lake, drawing attention from volcanologists and tourists alike. In 2017, a major eruption caused the lava lake to overflow, resulting in the formation of new lava flows that impacted the surrounding area. Increased volcanic gas emissions posed risks to nearby communities.

Fogo, Cape Verde. Fogo is a prominent stratovolcano located in a volcanic archipelago in the island nation Cape Verde, approximately 480 km off the west coast of Africa. The volcano has a geological history that spans about 150,000 years, with numerous eruptions shaping its landscape, including the formation of a large caldera. The explosive eruption of Fogo in 1995 devastated the town of Chã das Caldeiras and surrounding areas. This eruption resulted in the displacement of hundreds of residents and significant damage to infrastructure. The volcanic activity was characterized by the emission of ash and gases, as well as the formation of new lava domes. The most recent eruptive event occurred in 2015.

Mount Karisimbi, Rwanda. Mount Karisimbi is a stratovolcano located in the Virunga Mountain range, straddling the border between Rwanda and the Democratic Republic of Congo. It is part of the East African Rift system and is one of the eight major volcanoes in the Virunga massif. Karisimbi reaches an elevation of 4,507 m, making it the highest peak in Rwanda. Geologically, Mount Karisimbi has a history of volcanic activity that dates back approximately 1 million years, with its most recent eruptions occurring in the last few thousand years. The most notable eruptions happened around AD 620 and AD 800, producing volcanic ash and lava flows that contributed to the formation of its distinct summit caldera.

Mount Kenya, Kenya. Mount Kenya is located in central Kenya, standing as the country's highest mountain at approximately 5,199 m. Its geological history spans over 3 million years, with a large eruption occurring around 2.6 million years ago, contributing to the formation of its massive caldera and the surrounding volcanic terrain. The last known eruption took place approximately 300 years ago, around 1670, which produced lava flows and ash, though it was relatively minor compared to earlier activities. The mountain is a UNESCO World Heritage site.

Mount Kilimanjaro, Tanzania. Mount Kilimanjaro is a stratovolcano situated in Tanzania, and it is the continent's highest peak, standing at approximately 5,895 meters. Its geological history dates back about 1 million years and is characterized by three distinct volcanic cones: Kibo, Mawenzi, and Shira. Kilimanjaro's last significant eruptions occurred around 360,000 years ago, with the most recent activity being primarily from the Kibo cone.

Mount Longonot, Kenya. Mount Longonot is a stratovolcano located in the Great Rift Valley of Kenya, approximately 60 km northwest of Nairobi. It rises to an elevation of about 2,776 m and is known for its distinctive caldera. The volcano's geological history dates back approximately 1.2 million years, with its formation characterized by explosive eruptions and subsequent lava flow activity. An explosive eruption in 1863 formed the current caldera and produced ash deposits that covered the landscape. Since then, Mount Longonot has remained dormant, with no recorded eruptions in modern history.

Mount Meru, Tanzania. Mount Meru is an active stratovolcano located about 70 km northeast of Mount Kilimanjaro in Tanzania. It rises to an elevation of approximately 4,566 m and is part of the Eastern Arc Mountains. The geological history of Mount Meru spans over 250,000 years and is characterized by a series of eruptions that have shaped its distinctive cone and caldera. The last significant activity at Mount Meru occurred in 1910 and 1870. Both events are well documented in the tephrochronology record.

Mount Nyamuragira, Democratic Republic of Congo. Mount Nyamuragira is an active shield volcano located in the Virunga Mountains of the Democratic Republic of Congo, near the city of Goma. It stands at approximately 3,058 m and is one of Africa's most active volcanoes, with a geological history dating back about 10,000 years. The most significant eruption occurred in 1977, when a massive eruption produced an extensive lava flow that reached Lake Kivu, resulting in widespread destruction and displacement of communities. Large lava fountains and significant ash emissions characterized this eruption. Another important eruption, which took place in 2010, involved the formation of a new lava flow and a lava lake within Mount Nyamuragira's crater. Mount Nyamuragira is within Virunga National Park, a UNESCO World Heritage site.

Mount Nyiragongo, Democratic Republic of Congo. Mount Nyiragongo is an active stratovolcano located in the Virunga Mountains of the Democratic Republic of the Congo, near the city of Goma. Renowned for its persistent volcanic activity and one of the world's largest lava lakes, Nyiragongo reaches an elevation of about 3,470 m. A sudden eruption in January 2002 produced a large lava flow that devastated parts of Goma, destroying thousands of homes and displacing over 400,000 residents. This eruption was characterized by a fast-moving lava flow that reached the outskirts of the city, accompanied by significant gas emissions, which complicated evacuation efforts. Additionally, in 1977, one of the deadliest eruptions in the region took place, resulting in the deaths of approximately 70 people due to lava flows and toxic gas emissions.

Ol Doinyo Lengai, Tanzania. Ol Doinyo Lengai is an active stratovolcano located in the East African Rift Valley in northern Tanzania. Standing at approximately 2,878 m, it is known for being the only volcano on Earth that erupts natrocarbonatite lava, which is much cooler and more fluid than typical basaltic lava. The volcano's geological history dates back over 1 million years, with significant eruptions recorded in 1966, 1983–1984, 2007, and 2013.

Piton de la Fournaise, Réunion. Piton de la Fournaise is an active shield volcano located on the southeastern side of Réunion Island, a French territory in the Indian Ocean. It stands at an elevation of approximately 2,632 m and is one of the most active volcanoes in the world, erupting over 150 times since the first recorded eruption in 1640. Two notable eruptions occurred in 2007 and 2010, which were significant for their duration and the volume of lava produced, leading to the creation of new lava fields. These eruptions were characterized by spectacular lava fountains and the formation of extensive lava flows that traveled several kilometers from the summit. Piton de la Fournaise's eruptions are typically nonexplosive, characterized by the outpouring of basaltic lava that creates wide, gently sloping surfaces. Piton de la Fournaise is recognized as a UNESCO World Heritage site.

VOLCANOES OF ASIA

Avachinsky, Russia. Mount Avachinsky is a stratovolcano located on the Kamchatka Peninsula in Russia, standing at approximately 2,741 m. One of the most notable eruptions occurred in 2001, marked by explosive activity that produced significant ash plumes and pyroclastic flows, ranking it a 4 on the Volcanic Explosivity Index (VEI). Earlier recorded eruptions include significant activity in the 1950s and 1960s, characterized by frequent ash emissions and lava flows that have shaped the surrounding landscape.

Barren Island, India. Barren Island is an active volcano located in the Andaman Sea, part of India's Andaman and Nicobar Islands. It is the only confirmed active volcano in India and is characterized by a conical shape and a summit caldera that reaches an elevation of 354 m. The geological history of Barren Island includes several eruptions, with activity recorded since the late 18th century. Barren Island erupted in 2017, 2005, and 1991, preceded by significant seismic activity, which resulted in explosive events that produced ash plumes and lava flows. These eruptions have contributed to the ongoing geological processes shaping the island.

Bezymianny, Russia. Bezymianny is a stratovolcano located on the Kamchatka Peninsula in Russia that rises to approximately 2,882 m and is part of the larger group of volcanoes in the Kliuchevskoy group. The most impactful eruption of Bezymianny occurred in 1956, which was a catastrophic event that transformed the volcano's summit and marked a significant change in its eruptive history. This eruption produced a massive ash cloud that reached around 20 km into the atmosphere and generated pyroclastic flows that devastated the surrounding area. Since the 1956 eruption, Bezymianny has remained active, with frequent eruptions recorded in subsequent decades, including notable activity in 1970, 1980, and from 1990 to 1992.

Hunga Tonga–Hunga Ha'apai, Tonga. Hunga Tonga is a volcanic island located in the South Pacific Ocean, part of the Kingdom of Tonga. This underwater volcano is known for its significant geological activity and the formation of two islands, Hunga Tonga and Hunga Ha'apai, which emerged as a result of past eruptions. The most notable eruption occurred on January 15, 2022, when a massive underwater volcanic explosion produced one of the most significant eruptions in recent history. The 2022 eruption generated a powerful shockwave, a towering plume of ash and steam that rose about 58 km into the stratosphere and was heard as far away as Alaska. This eruption triggered widespread tsunamis that affected Tonga and reached distant shores, causing damage to infrastructure, homes, and communications in the region. The explosion also released a significant amount of volcanic gas into the atmosphere, raising concerns about its potential climatic impacts. Prior to the 2022 event, Hunga Tonga-Hunga Ha'apai had experienced eruptions in 2009 and 2015, which were less explosive but contributed to the ongoing geological activity of the volcano.

Krakatoa, Indonesia. Krakatoa, located in the Sunda Strait between the islands of Java and Sumatra in Indonesia, is one of the most infamous volcanoes in history due to its catastrophic eruptions. The volcanic complex includes several islands, with the most famous being the original Krakatoa and Anak Krakatau (Child of Krakatoa), which emerged after the 1883 eruption. The most notable eruption occurred on August 26–27, 1883, and was one of the deadliest volcanic events in recorded history. This explosive eruption produced massive ash clouds, loud sonic booms, and tsunamis that devastated coastal communities, resulting in over 36,000 fatalities. The eruption altered the global climate, leading to a significant drop in temperatures and spectacular sunsets due to ash particles in the atmosphere.

Mount Agung, Indonesia. Mount Agung is a stratovolcano located on the island of Bali, Indonesia, and is the highest peak in Bali at approximately 3,031 m. Its geological history spans over 20,000 years, marked by several significant eruptions, including the 1963–1964 event, one of the most powerful eruptions in Bali's history. This eruption produced a massive ash cloud that reached up to 20 km into the atmosphere, resulting in pyroclastic flows and lava flows that devastated nearby villages, causing over 1,000 fatalities. More recently,

Mount Agung erupted again in 2017–2018, with increased seismic activity leading to ash emissions and lava flows that prompted the evacuation of thousands of residents.

Mount Anak Krakatau, Indonesia. Following the catastrophic 1883 eruption of Krakatoa, volcanic activity continued intermittently, leading to the formation of Anak Krakatau, which began to emerge in 1927. This new island has experienced several eruptions, including notable events in 1994 and 2018, the latter of which triggered a deadly tsunami.

Mount Aso, Japan. Mount Aso is an active stratovolcano located on the island of Kyushu, Japan, and is one of the largest volcanoes in the country. It features a complex caldera measuring about 25 km in diameter, formed through multiple explosive eruptions over the past 300,000 years. The volcano is characterized by its central cone, which includes several peaks, with Nakadake being the highest at 1,506 m. Mount Aso erupted two times in recent history: 1958 and 2016. Both prompted an evacuation, and the 2016 event produced an ash cloud that disrupted air travel.

Mount Fuji, Japan. Mount Fuji, an iconic stratovolcano located on Honshu Island, Japan, stands at 3,776 m and is recognized as the highest peak in Japan. Its geological history spans over 100,000 years and is characterized by volcanic activity that has shaped its symmetrical cone and surrounding landscape. The most notable eruption of Mount Fuji occurred in December 1707, known as the Hōei eruption. This explosive event produced a significant ash cloud that spread over vast areas, affecting air quality and agriculture in nearby regions, including Edo (modern-day Tokyo). The eruption lasted for about 2 weeks and resulted in the deposition of an estimated 800,000 tons of ash, which covered surrounding towns and farmland. Prior to the Hōei eruption, Fuji experienced several eruptions, with records indicating activity approximately every few hundred years. The last recorded eruption occurred in 1707, and since then, Mount Fuji has remained dormant, though it is still monitored for potential volcanic activity. The volcano is a UNESCO World Heritage site celebrated for its cultural significance and breathtaking scenery.

Mount Iwo-Jima, Japan. Mount Iwo-Jima, also known as Iwo To, is an active stratovolcano located in the Philippine Sea, about 1,200 km south of Tokyo, Japan. The island's volcanic activity has been documented for centuries, with a geological history characterized by explosive eruptions and the formation of a rugged landscape. One of the most significant eruptions occurred in 1986, when Iwo Jima experienced a series of explosive events that produced ash plumes and pyroclastic flows. This eruption prompted the evacuation of residents and heightened monitoring of the volcano's activity. Prior to the 1986 eruption, Iwo-Jima had notable eruptions in the late 19th century, including an explosive event in 1889 that resulted in the formation of new craters. The island's volcanic activity remains a subject of interest, as it is part of the Izu-Bonin arc, a region known for its tectonic activity. Mount Iwo-Jima is also historically significant, as it was the site of fierce battles during World War II.

Mount Merapi, Indonesia. Mount Merapi is a stratovolcano located on the island of Java, Indonesia, and is one of the most active volcanoes in the world. Rising to about 2,930 m, it has a long and turbulent geological history characterized by frequent eruptions, with significant activity recorded for over 1,500 years. One of the most notable eruptions occurred in 2010, and was particularly destructive. This eruption produced pyroclastic flows and ash plumes that reached heights of 7 km, forcing the evacuation of thousands of residents and resulting in over 300 fatalities. Another significant eruption took place in 2006, when increased seismic activity led to the expulsion of ash and smoke, prompting evacuations and heightened alert levels. Merapi's eruptions are typically characterized by both explosive and effusive phases, often involving the emission of lava flows and ash clouds.

Mount Ontake, Japan. Mount Ontake is an active stratovolcano located in central Japan, standing at 3,067 m. It is the second-highest volcano in Japan and is characterized by numerous eruptions, with activity dating back over 100,000 years. The last eruption occurred on September 27, 2014, when a sudden explosive event released a large ash cloud, resulting in the deaths of 63 hikers caught in the eruption. This eruption was particularly impactful due to its unexpected nature, as it followed a period of relative quiet and was not preceded by significant warning signs. The ashfall affected nearby communities, disrupting air quality and leading to evacuation orders. Earlier eruptions include 1979 and 1980, which produced ash and volcanic gas emissions, although they were less destructive than the 2014 event.

Mount Paektu, China/North Korea. Mount Paektu, also known as Changbai Mountain, is a stratovolcano located on the border between China and North Korea. Standing at approximately 2,744 m, it is the highest mountain in the Korean Peninsula and holds significant cultural and historical importance. The volcano has a complex geological history, with eruptions dating back over 2,000 years. The eruption of AD 946, known as the Millennium Eruption, is one of the most significant volcanic events in recorded history. This explosive eruption generated a massive ash cloud that affected large areas of Asia, resulting in significant climatic changes and agricultural impacts. The eruption also formed the caldera at the summit, which is now filled with a crater lake called Lake Chon. Other significant eruptions include those in 1668 and 1903.

Mount Pinatubo, Philippines. Mount Pinatubo, a large stratovolcano, is located on the island of Luzon in the Philippines and is known for its catastrophic eruption in June 1991, one of the most significant volcanic events of the 20th century. The volcano has a complex geological history, with its most significant eruptions occurring primarily in the last 500,000 years. The 1991 eruption followed a series of minor eruptions that began in April of that year, culminating in a massive explosion on June 15. This eruption produced a colossal ash cloud that reached approximately 35 km into the atmosphere, resulting in widespread ashfall across the Philippines and as far away as the United States. The eruption also triggered pyroclastic flows and lahar events, causing extensive destruction and displacing around 250,000 people, resulting in significant damage to the surrounding areas, including the nearby Clark Air Base. Prior to 1991, Mount Pinatubo had been relatively quiet for hundreds of years, with the last major eruption recorded in 1600. The aftermath of the 1991 eruption included profound climatic effects, leading to a temporary global temperature drop due to the release of volcanic aerosols into the atmosphere.

Mount Sakurajima, Japan. Mount Sakurajima is an active stratovolcano located in Kagoshima Prefecture, Japan, on the island of Kyushu. Mount Sakurajima formed initially as a single island, but significant eruptions in the 20th century led to the formation of a volcanic cone that is now connected to the Osumi Peninsula. Mount Sakurajima has continued to erupt regularly, with notable eruptions recorded in 1946, 1955, and through the 2000s, including a series of eruptions starting in 2009 that have led to increased volcanic activity and ash emissions. The volcano is closely monitored due to its proximity to populated areas, with ongoing efforts to assess hazards and ensure public safety.

Mount Tambora, Indonesia. Mount Tambora is an active stratovolcano located on the island of Sumbawa in Indonesia. It is most famous for its catastrophic eruption in 1815, one of the most powerful volcanic eruptions in recorded history. The 1815 eruption produced a massive ash cloud that reached the stratosphere, resulting in significant cooling and abnormal weather patterns worldwide. Local populations suffered severe consequences, with estimates of tens of thousands of deaths due to the eruption and its aftermath, including famine and disease. The eruption reduced the height of Mount Tambora from about 4,300 m to approximately 2,850 m. Mount Tambora is recognized as a UNESCO World Heritage site.

Mount Unzen, Japan. Mount Unzen is an active stratovolcano located in Nagasaki Prefecture on the island of Kyushu,

Japan. The volcano is known for its steep slopes and striking peaks, including the prominent Fugen-dake. Eruptions from 1990 to 1995 were marked by a series of explosive events that produced pyroclastic flows and ash clouds. The last eruption was particularly devastating, resulting in the loss of 43 lives, including scientists and local residents. Earlier eruptions, such as the 1792 eruption, were also destructive, causing a massive landslide that triggered a tsunami, resulting in approximately 15,000 fatalities.

Taal, Philippines. Taal Volcano is a stratovolcano located on Luzon Island in the Philippines, about 70 km south of Manila. It is one of the most active volcanoes in the country, with a history of eruptions dating back over 400 years. Taal is unique because it is situated within an enormous caldera formed by previous explosive eruptions, hosting a smaller island (Volcano Island) with its own crater lake. The most recent eruption, in January 2020, was marked by explosive activity that produced ash plumes reaching up to 15 km into the atmosphere. This eruption prompted mass evacuations of nearby residents, impacting thousands of people and causing significant ashfall in surrounding areas, including Metro Manila. Earlier significant eruptions include those in 1911, which resulted in over 1,300 fatalities due to pyroclastic flows and ashfall, and the 1977 eruption, which produced a substantial ash cloud.

Ulawun, Papua New Guinea. Ulawun is an active stratovolcano located on the island of New Britain in Papua New Guinea, standing at approximately 2,334 m. Known for its symmetrical cone and frequent volcanic activity, Ulawun has a geological history that spans thousands of years, with its most significant eruptions documented since the 18th century. Recent eruptions include a series of events in 2013 and 2016, which featured explosive eruptions that prompted evacuations and disrupted air travel. The 2016 event produced a large ash cloud that affected visibility and air quality in the region.

VOLCANO OF AUSTRALIA

Big Ben Volcano, Australia. Big Ben is an active volcano located on Heard Island, which is part of Australia's external territory in the Southern Indian Ocean. This stratovolcano rises to about 2,745 m and is the highest peak in the Australian Antarctic Territory. Big Ben is known for its distinctive conical shape and is surrounded by glaciers. Big Ben's eruptions typically involve both explosive and effusive phases, as was documented in 1980 and 1981.

VOLCANOES OF ANTARCTICA

Mount Erebus, Ross Island. Mount Erebus, located in Antarctica on Ross Island, is the world's southernmost volcano. Standing at approximately 3,794 m, it is recognized for its persistent lava lake, one of the few permanent lava lakes in the world. Its activity is generated by the Erebus Hot Spot, which triggered significant phreatic eruptions in 1993 and 1972.

Port Foster, Deception Island. Port Foster is a large caldera located on Deception Island, part of the South Shetland Islands in Antarctica. A series of explosive events formed the caldera. Deception Island is unique due to its horseshoe shape, created by the collapse of the volcanic structure, and it features a sheltered harbor that has historically been used by whalers and researchers. An eruption in 1967 led to the evacuation of a nearby research station.

BIBLIOGRAPHY

Acocella, V. 2021. "Calderas." In *Volcano-Tectonic Processes: Advances in Volcanology*, edited by K. Nemeth, 13–25. Springer.

Ambrose, S.H. 2003. "Did the super eruption of Toba cause a human population bottleneck? Reply to Gathrone-Hardy and Harcourt-Smith." *Human Evolution* 45 (3): 231–237.

Axelsson, G. 2024. "The future of geothermal energy." In *Living with Climate Change*, 1st ed., edited by E. Tolsley, 397–422. Elsevier.

Begét, J.E. 2000. "Volcanic tsunamis." In *The Encyclopedia of Volcanoes*, 2nd ed., edited by Haraldur Sigurdsson, 105. Academic Press.

Bindeman, I.N. 2006. "The secrets of supervolcanoes." *Scientific American* 294 (6): 36–43.

Bindschadler, D.L. 1995. "Magellan: A new view of Venus' geology and geophysics." *American Geophysical Union* 19 (3): 35–47.

Bond, D.P., and S.E. Grasby. 2017. "On the cause of mass extinctions." *Palaeogeography, Palaeoclimatology, Palaeoecology* 478: 3–29.

Boreham, A., K. Cashman, A. Rust, and A. Hoskuldsson. 2018. "Linking lava flow morphology, water availability and rootless cone formation on the Younger Laxa Laa, NE Iceland." *Journal of Volcanology and Geothermal Research* 364 (15): 1–19.

Branney, M.J., B. Bonnichsen, G.D.M. Andrews, B. Ellis, T.L. Barry, and M. McCurry. 2008. "Snake River (SR)-type" volcanism at the Yellowstone hotspot track: Distinctive products from unusual, high-temperature silicic supereruptions." *Bulletin of Volcanology* 70: 293–314.

Calder, E.S., Y. Lavallée, J.E. Kendrick, and M. Bernstein. 2015. "Lava dome eruptions." In *The Encyclopedia of Volcanoes*, 2nd ed., edited by Haraldur Sigurdsson, 343–362. Academic Press.

Carey, R.J., B.F. Houghton, and T. Thordarson. 2010. "Tephra dispersal and eruption dynamics of wet and dry phases of the 1875 eruption of Askja Volcano, Iceland." *Bulletin of Volcanology* 72 (3): 259–278.

Carrivick, J.L., J.R. Andrew, and S.T. Fiona. 2004. "Geomorphological evidence for jökulhlaups from Kverkfjöll volcano, Iceland." *Geomorphology* 63 (1–2): 81–102.

Decker, R.W., and B. Decker. 2005. *Volcanoes*. 4th ed. W.H. Freeman & Company.

Deligne, N.I., and H. Sigurdsson. 2015. "Global rates of volcanism and volcanic episodes." In *The Encyclopedia of Volcanoes*, 3rd ed., edited by Haraldur Sigurdsson, 349–352. Academic Press.

DeSilva, S., and S. Self. 2022. "Extreme in volcanology: The case for the term 'supervolcano.'" *Frontiers in Earth Science* 10: 1–7.

Dull, R.A., J.R. Southon, S. Kutterolf, K.J. Anchukaitis, A. Freundt, D.B. Wahl, P. Sheets, P. Amaroli, W. Hernandez, M.C. Wiemann, and C. Oppenheimer. 2019. "Radiocarbon and geological evidence reveal Ilopango volcano as source of the colossal 'mystery' eruption of 539/40 CE." *Quaternary Science Reviews* 222: 67–73.

Dull, R.A., J.R. Southon, and P. Sheets. 2001. "Ecology and culture: A reassessment of the Volcan Ilopango TBJ eruption in the southern Maya realm." *Latin American Antiquity* 12 (1): 25–44.

Earle, S. 2019. *Physical Geology*. Pressbooks.

Encyclopedia Britannica. 2012. Volcanic island evolution. https://www.britannica.com.

Encyclopedia Britannica, Inc. 2025. Volcanic island formation (Illustration). *Encyclopedia Britannica*. http://www.britannica.com/science/volcanicislands.

Ernst, E., E. Grosfils, and D. Mege. 2001. "Giant dike swarms: Earth, Venus, and Mars." *Annual Review of Earth and Planetary Sciences* 29: 489–435.

Fink, J.H., and S.N. Anderson. 2015. "Lava domes and coulees." In *The Encyclopedia of Volcanoes*, 2nd ed., edited by Haraldur Sigurdsson, 308. Academic Press.

Fink, J., and S. Anderson. 2023. "Lessons learned from the 1980–1986 eruption of the Mount St. Helens composite lava dome." *Bulletin of Volcanology* 85 (35): 34–41.

Fiocco, G., D. Fuà, and G. Visconti. 1996. *The Mount Pinatubo Eruption: Effects on the Atmosphere and Climate.* Springer.

Gathorne-Hardy, F.J., and W.E.H. Harcourt-Smith. 2003. "The super-eruption of Toba: Did it cause a human bottleneck? *Journal of Human Evolution* 45 (3): 227–230.

Ge, Y., and X. Gao. 2003. "Understanding the overestimated impact of the Toba volcanic super-eruption on global environments and ancient hominins." *Journal of Human Evolution* 45 (3): 231–237.

Geissler, P. 2015. "Cryovolcanism in the outer solar system." In *The Encyclopedia of Volcanoes*, 2nd ed., edited by Haraldur Sigurdsson, 785–800. Academic Press.

Genareau, K., J.B. Wardman, T.M. Wilson, S.R. McNutt, and P. Izbekov. 2015. "Lighting-induced volcanic spherules." *Geology* 43 (4): 319–322.

Global Volcanism Program. 2024. *Volcanoes of the World*. Version 5.2.3. Distributed by Smithsonian Institution, compiled by Venzke, E. https://doi.org/10.5479/si.GVP.VOTW5-2024.5.2 (accessed September 20, 2024).

Hartley, M.E., and T. Thordarson. 2013. "The 1874–1876 volcano-tectonic episode at Askja, North Iceland: Lateral flow revisited." *Geochemistry, Geophysics, Geosystems* 14 (7): 29–31.

Helffrich, G.R., and B.J. Wood. 2001. "The Earth's mantle." *Nature* 412 (6846): 501–507.

Huang, C.Y., M.X. Zhao, C.C. Wang, and G.J. Wei. 2001. "Cooling of the South China Sea by the Toba eruption and correlation with other climate proxies ~71,000 years ago." *Geophysical Research Letters* 28: 3915–3918.

Huang, H.H., L. Fan-Chi, B. Schmandt, J. Farrell, R.B. Smith, and C. Victor. 2015. "The Yellowstone magmatic system from the mantle plume to the upper crust." *Science* 348 (6236): 773–776.

Ivanov, A.V., H. Huayiu, Y. Liekun, V. Viktor, A.Y. Ryabov, S.V. Palesskii, and I.V. Nikolaeva. 2013. "Siberian Traps large igneous province: Evidence for two flood basalt pulses around the Permo-Triassic boundary and in the Middle Triassic, and contemporaneous granitic magmatism." *Earth-Science Reviews* 122: 58–76.

Jakobsson, S. P., and M.T. Gudmundsson. 2009. "Subglacial and intraglacial volcanic formations in Iceland." *Jökull* 58: 179–196.

Jakobsson, S. P., and G. L. Johnson. 2012. "Intraglacial volcanism in the Western Volcanic Zone, Iceland." *Bulletin of Volcanology* 74 (5): 1141–1160.

John, S.E., O. J. Rouxel, P.R. Craddock, A.M. Engwall, and E.A. Boyle. 2008. "Zinc stable isotopes in seafloor hydrothermal vent fluids and chimneys." *Earth and Planetary Science Letters* 10: 17–28.

Jovanelly, T. 2020. *Iceland: Tectonics, Volcanics, and Glacial Features*. Wiley-Blackwell.

Kargel, J.S. 1994. "Cryovolcanism on the icy satellites." *Earth Moon Planet* 67: 101–113.

Kauahikaua, J., D.R. Sherrod, C. Katharine, C. Heliker, K. Hon, T.N. Mattox, and J.A. Johnson. 2003. "Hawaiian lava-flow dynamics during the Pu'u Oo-Ku paianaha eruption: A tale of two decades." US Geological Survey Professional Paper 1676.

Kilburn, C.R. 2000. "Lava flows and flow fields." In *The Encyclopedia of Volcanoes*, 2nd ed., edited by Haraldur Sigurdsson, 343–362. Academic Press.

Kogure, T., R. Sueyoshim, O. Hiroto, S. Yoshikazu, S. Ki-Cheol, and A. Yutaka. 2022. "Formation processes of tafone on pyroclastic rock surfaces with hydrothermal alteration on the Isotake coast, Shimane, Japan." *Geomorphology* 398: 54–67.

Koike, M., and H. Nakamichi. 2021. "Dike inflation process beneath Sakurajima volcano, Japan, during the earthquake swarm of August 15 2015." *Frontiers in Earth Science* 8: 30–35.

Lane, C. S., T. Ben., T. Chorn, and T.C. Johnson. 2013. "Reply to Roberts et al.: A sub decadal record of paleoclimate around the youngest Toba Tuff in Lake Malawi." *Proceedings of the National Academy of Sciences* 110 (330): E3048–E3048.

Lin, J., A. Svensson, C.S. Hvidberg, J. Lohmann, S. Kristiansen, D. Dahl-Jensen, J. P. Steffensen, S.O. Rasmussen, E. Cook, H. A. Kjaer, B.M. Vinther, H. Fischer, T. Stocker, M. Sigl, M. Bigler, M. Severi, R. Traversi, and R. Mulvaney. 2022. "Magnitude, frequency and climate forcing of global volcanism during the last glacial period as seen in Greenland and Antarctic ice cores (60–9 ka)." *Journal Climate of the Past* 18: 485–506.

Matter, M., W.S. Broecker, S.R. Gislason, E. Gunnlaugsson, E.H. Oeikers, M. Slute, H. Sigurdardottir, A. Stefansson, H.A. Alfredsson, E.S. Aradottir, G. Axelsson, B. Sigfusson, and D. Wolff-Boenisch. 2011. "The CarbFix Pilot Project—Storing carbon dioxide in basalt." *Energy Procedia* 4: 5579–5585.

May, R.J. 1979. "Thermoluminescence of the Hawaiian island basalts." United States Geological Survey Professional Paper 1095, p. 52.

McCormick, M.P., L.W. Thomason, and C.R. Trepte. 1995. "Atmospheric effects of the Mt. Pinatubo eruption." *Nature* 373: 399–404.

McNutt, S.R. and E.R. Williams. 2010. "Volcanic lightening: global observations and constraints on source mechanisms." *Bulletin of Volcanology* 72 (10): 1153–1167.

Meredith, E. S., S.F. Jenkins, and J.L. Hayes. 2022. "Damage assessment for the 2018 lower East Rift Zone lava flows of Kilauea volcano, Hawaii." *Bulletin of Volcanology* 84 (65): 17–23.

Moore, J. G. 1992. "Structure and eruptive mechanisms at Surtsey Volcano, Iceland." *Geological Magazine* 122 (6): 649–661.

Mora-Amador, R.A., D. Rouwet, P. Vargas, and C. Oppenheimer. 2019. "The extraordinary sulfur volcanism of Poás from 1828–2018." In *Poas Volcano: The Pulsing Heart of Central America Volcanic Zone*, 1st ed., edited by Franco Tassi, Orlando Vaselli, and Raul Alberto Mora Amador, 45–78. Springer.

Muehlberger, W. R., S. J. Muehlberger, and L. G. Price. 2005. *High Plains of Northeastern New Mexico: A Guide to Geology and Culture*. Scenic Trip Series 19. New Mexico Bureau of Geology and Mineral Resources.

National Aeronautics and Space Administration, Jet Propulsion Laboratory. 2025. NASA Juno mission spots powerful volcanic activity on IO. https://www.nasa.gov/missions/juno/nasa-juno-mission-spots-most-powerful-volcanic-activity-on-io-to-date/.

National Oceanic and Atmospheric Administration. 2023, December 20. "A volcanic eruption sent enough water vapor into the stratosphere to cause rapid change in chemistry." https://research.noaa.gov/2023/12/20/hunga-tonga-2022-eruption/.

Newhall, C.G., and S. Self. 1982. "The volcanic explosivity index (VEI): An estimate of explosive magnitude for historical volcanism." *Journal of Geophysical Research* 87 (C2): 1231–1238.

Nobuo, G. 2008. "Piston to funnel—successive growth of a collapsed caldera during the Miyakejima 2000 eruption." *Collapse Calderas Workshop*. IOP Conf. Series: Earth and Environmental Science. IOP Publishing. https//doi.org/10.1088/1755-1307/3/1/012022.

Nomade S, D. Genty, R. Sasco, V. Scao, V. Féruglio, and D. Baffier. 2016. "A 36,000-year-old volcanic eruption depicted in the Chauvet-Pont d'Arc cave (Ardèche, France)." *PLoS ONE* 11 (1): 114–118.

Ohtani, E, 2020. "The role of water in the Earth's mantle." *National Science Review* 7 (1): 224–232.

Oppenheimer, C. 2002. "Limited global change due to the largest known Quaternary eruption, Toba ~74 kyr BP." *Quaternary Science Review* 21: 1593–1609.

Osipov, S., G. Stenchikov, and K. Tsigaridis. 2021. "The Toba supervolcano eruption caused severe tropical stratospheric ozone depletion." *Communications Earth & Environment* 2 (71): 12–16.

Pierson T.C., N.J. Wood, and C.L. Driedger. 2014. "Reducing risk from lahar hazards: Concepts, case studies, and roles for scientists." *Journal of Applied Volcanology* 3 (16): 1–25.

Ping, C. 2000. "Volcanic soils." In *The Encyclopedia of Volcanoes*, 3rd ed., edited by Haraldur Sigurdsson, 349–352. Academic Press.

Plunkett, G., M. Sigl, J.R. McConnell, J.R. Pilcher, and N.J. Chellman. 2023. "The significance of volcanic ash in Greenland ice cores during the Common Era." *Quaternary Science Reviews* 301: 18–36.

Prothero, D. 2018. *When Humans Nearly Vanished: The Catastrophic Explosion of the Toba Volcano*. Smithsonian Books.

Rampino, M.R., and S. Self. 2000. "Volcanism and biotic extinction." In *The Encyclopedia of Volcanoes*, 2nd ed., edited by Haraldur Sigurdsson, 1083–1091. Academic Press.

Raymond, A. P. Duraiswami, T. Gadpallu, and N. Shaikh. 2014. "Pahoehoe—a'a transitions in the lava flow fields of the western Deccan Traps, India—Implications for emplacement dynamics, flood basalt architecture and volcanic stratigraphy." *Journal of Asian Earth Sciences* 84: 146–166.

Renne, P.R., and A.R. Basu. 1991. "Rapid eruption of the Siberian Traps flood basalts at the Permo-Triassic boundary." *Science* 253 (5016): 176–179.

Romagnoli, C., and S.P. Jakobsson. 2015. "Post-eruptive morphological evolution of island volcanoes: Surtsey as a modern case study." *Geomorphology* 250: 384–396.

Rose, I.W., and A.C. Chesner. 1990. "Worldwide dispersal of ash and gases from earth's largest known eruption: Toba, Sumatra, 75 ka." *Palaeogeography and Palaeoecology* 89: 269–275.

Ruddiman, W.F. 2014. *Earth's Climate: Past and Future*. 2nd ed. Freeman and Company.

Rymer, H. 2000. "Volcanic hazards." In *The Encyclopedia of Volcanoes*, 2nd ed., edited by Haraldur Sigurdsson, 911. Academic Press.

Sarikaya, M. A., A. Ciner, and M. Zreda. 2015. "Fairy chimney erosion rates on Cappadocia ignimbrites, Turkey: Insights from cosmogenic nuclides." *Geomorphology* 234: 182–191.

Sauders, A., and M. Reichow. 2009. "The Siberian Traps and the End-Permian mass extinction: A critical review." *Chinese Science Bulletin* 54: 20–37.

Showstack, R. 2011. "Nailing Pompeii." *EOS American Geophysical Union* 78 (36): 382–383.

Simkin, T. and L. Siebert. 2000. "Earth's volcanoes and eruptions." In *The Encyclopedia of Volcanoes*, 2nd ed., edited by Haraldur Sigurdsson, 265–272. Academic Pres.

Spijkerboer, R., E. Turhan, A. Roos, M. Billi, S. Vargas-Payers, S. Opazo, and M. Armiero. 2022. "Out of steam? A social science and humanities research agenda for geothermal energy." *Energy Resource and Social Science* 92 (1–10): 67–89.

Spry, A. 1962. "The origin of columnar jointing, particularly in basalt flows." *Journal of Geological Society of Australia* 8 (2): 191–216.

Stovall, W.K., C.L. Driedger, E.G Westby, and L.M. Faust. 2019. "Living with volcano hazards." U.S. Geological Survey Fact Sheet Report 2018-3075.

Thorarinsson, S. 1964. *Eldur í Öskju: Askja on fire*. Reykjavík: Almenna Bókafélagið.

Thorarinsson, S. 1966. "The median zone of Iceland." *The World Rift System* 66: 187.

Thorarinsson, S. 1968. "The last phases of the Surtsey eruption." *Natturufraedingurinn* 38: 113–135.

Thordarson, T., and S. Self. 2003. "Atmospheric and environmental effects of the 1783–1784 Laki eruption: A review and reassessment." *Journal of Geophysical Research* 108 (D1): 80–96.

Tómasson, H. 1996. "The jökulhlaup from Katla in 1918." *Annuals of Glaciology* 22: 249–254.

United States Geological Survey, Volcanics Hazard Program. 2020. https://www.usgs.gov/programs/VHP/vhp-uses-monitoring-data-and-volcanic-history-forecast-eruptions (accessed November 30, 2024).

Vallance, J.W. 2000. "Lahars." In *The Encyclopedia of Volcanoes*, 2nd ed., edited by Haraldur Sigurdsson, 602–603. Academic Press.

Van Wyk de Vries, B., D. Karatsch, and C. Gouard. 2022. "Inverted volcanic relief: Its importance in illustrating geologic change and its geoheritage potential." *Journal of Geoheritage and Parks* 10 (1): 47–83.

Wang, B.W. 2024. "Returned samples indicate volcanism on the Moon 120 million years ago." *Science* 385 (6713): 1077–1080.

White, J.D., J.L. Smellie, and D.A. Clague. 2003. "Introduction: A deductive outline and topical overview of subaqueous explosive-volcanism." In *Explosive Subaqueous Volcanism*, 1st ed., edited by J.D.L White, 1–23. American Geophysical Union Monograph, Washington D.C.

Wu Y, D. Chu, and J. Tong. 2021. "Six-fold increase of atmospheric pCO2 during the Permian-Triassic mass extinction." *Nature Communications* 12: article 2137.

Zaki, A.S. 2016. "Inverted topography in the southeast part of the western desert of Egypt." *Journal of African Earth Science* 121: 56–61.

Zielinski G.A., P.A. Mayewski, L.D. Meeker, S. Whitlow, M.S. Twickler, and K. Taylor. 1996. "Potential atmospheric impact of the Toba mega-eruption ~71,000 years ago." *Geophysical Research Letters* 23: 837–840.

INDEX

For the benefit of digital users, indexed terms that span two pages (e.g., 52–53) may, on occasion, appear on only one of those pages.

Figures are indicated by *f* following the page number

A

I

N